Welded joint design

Welded joint design

Third edition

John Hicks

ABINGTON PUBLISHING

Woodhead Publishing Limited in association with The Welding Institute
Cambridge England

First published by Granada Publishing in Crosby Lockwood Staples 1979
Second edition published by BSP Professional Books 1987
Second edition reprinted by Abington Publishing 1997
Third edition published 1999 by Abington Publishing, Woodhead Publishing
Limited, Abington Hall, Granta Park, Great Abington, Cambridge CB21 6AH,
England
www.woodheadpublishing.com; www.woodheadpublishingonline.com
Reprinted 2012

British Library Cataloguing in Publication Data
A catalogue record for this book is available from the British Library.

ISBN 978-1-85573-386-2 (book)
ISBN 978-1-85573-898-0 (e-book)

Contents

Foreword
by
Dr A A Denton
CBE MA PhD DIC HonDSc FEng FIMechE FRINA
Past President of the Institution of Mechanical Engineers

Welding occupies a key position in the manufacture of almost every product which we use; the integrity of the welded joints is vital whether it be a microcircuit or an offshore installation. At the outset of my career I worked on the development of the tubular electrode and automatic submerged arc welding machines for shipyard use. It was a time when arc welding processes were well established and widely used but by today's standards the parent and weld metal properties and the stability of the processes were somewhat uncertain.

Although the technologies of welding have now matured their proper application and exploitation still calls for specialist input both in design and in fabrication. The qualified welding engineer is the person who can attend to these specialised requirements and to do this effectively he needs to understand the basis of the design and the significance of the welded details. At the same time it is important that the practising designer acquires a basic knowledge of the relevant aspects of welding to be able to execute satisfactory designs and, equally important, to know when to seek the input of a qualified welding engineer.

This book, whose author has 35 years' experience in design for welding, offers a straightforward explanation of the basis of design of welded structures and of the constraints which welding may impose on the design. It is therefore a work which should be seen to be of value to both the welding engineer and the design engineer.

Preface

The profession of the welding engineer is multidisciplinary; many of those coming into it have been educated and trained in electrical engineering, mechanical engineering, production engineering, metallurgy or in other disciplines but do not have a background in the design and construction of welded structures and plant. I have written this book mainly for such engineers and in so doing I have drawn on my observations as a moderator for the subject of Design and Construction in the European Welding Federation Examinations for Welding Engineer, Technologist and Specialist conducted by TWI in the United Kingdom. Candidates for these examinations are not expected to be able to undertake design work themselves but are expected to have acquired an understanding of the design processes which lead to their being presented with specifications and drawings for fabrications. The book explains the basis of typical design requirements and decisions which lead to the features characteristic of welded structures and plant. I am indebted to Mr Mark Cozens and Mr Ken Verrill, both of TWI, for their helpful advice on the content and presentation of this book.

The welding engineer in various industries will come across designs of fabricated structures for buildings, bridges, ships, railway rolling stock, road vehicles, dock gates, cranes and other mechanical handling equipment, pressure or vacuum vessels, pipework and pipelines and all types of machinery. The significance of the word *structures* is that it indicates objects or parts of objects which have to carry, or resist, loads. The loads may be the deadweight of the structure itself, something which it carries, a reaction to acceleration, environmental loads, pressure or thermal expansion, etc. .The safe performance of a structure relies on materials and methods of fabrication which can respond to the explicit or implicit design requirements. It is important that the welding engineer has the opportunity of making his specialist input to the design process, and an understanding of the basis of the design will help that contribution to be most effective.

The derivation of many of the relationships which explain the behaviour of solids rests on rigorous mathematical analyses whose basis or results may be tempered by assumptions or simplifications in order to arrive at a practical solution. On the other hand, some design criteria are merely empirical values or relations and have no theoretical basis. This book does not attempt to justify, nor in some cases explain in detail the derivation of, the approaches which are used and asks the reader to accept what is written as what is practised to help understand the processes which lead to the creation of a design for a fabricated product. As a result, the book is not intended to help the reader to become a structural engineer; nonetheless, practising engineers in many industries will find within this book an understanding of the scope and need for specialist welding input to the design of welded structures and plant. I have found in my own experience that what at first sight may seem to be fearsomely complicated methods are quickly understood by performing calculations using the methods. I have for this purpose put some examples in the text. I have also put some problems at the end of certain chapters with the answers at the end of the book but these are not to be thought of as samples of examination questions. A reader wishing to study any matter in depth is referred to the bibliography at the end of the book.

The process of design of most fabricated products, even at what in some industries is called the conceptual stage, rarely starts with an appeal to first principles. Much design is a reiteration of previous work meeting new functional requirements. Many of the design criteria are based entirely on the requirements of a standard specification and there are few products these days which do not have to recognise some industry, national, regional or international standard specifications. Over recent years there has been an enormous increase in the production of new standard specifications in the field of welding and its applications as well as revisions to existing ones. In a large part this has been due to the moves to construct a European based collection of standard specifications through the European Committee for Standardisation (CEN) which have in many cases been combined with international interests through the International Organisation for Standardisation (ISO). Standards provide a useful basis for national, regional or worldwide compatibility and interchangeability for artefacts and an agreed approach to achieving comparability in design and fabrication methods with the inherent contribution to safety. The apparently comprehensive nature of many modern standards can give them an air of absolute technical and scientific authority; however, they represent only a consensus of views or practices of self-selected groups and can never be vehicles for the use or development of novel concepts or recently acquired knowledge. They should not be used without an understanding of their engineering or scientific basis or the limitations to the scope of their application. It is therefore imperative that the user has a sound understanding of the basis of the technology so as to be capable of reaching a view on the validity and scope of a standard and to make decisions on matters falling outside the scope of that standard. There is often also the more mundane position where a standard nominated in a contract is inappropriate for the work and the knowledgeable welding engineer can ensure that steps are taken to prevent costly and inappropriate compromises having to be made just to meet what is in effect an irrelevant, or at best ambiguous, requirement.

I am aware that many readers may find it a convenience to have included in a book such as this some of the data which appear in standards, e.g. weld symbols, fatigue life data and allowable stresses, etc. The rate at which standards are currently being developed and amended could result in such material rapidly becoming out of date and despite the growth of international standards there are still many national and regional standards in use which differ from each other. For these reasons I have decided not to include direct quotations from standards except as examples.

John Hicks

Fundamentals of the strength of materials

Normal stress

Materials change in length when they are put under normal (or direct) stress which can be either tensile or compressive stress. An elastic material is one in which the change in length is proportional to the stress developed in it and also one in which the material will return to its original length after the stress is removed. Many metals behave in an elastic manner up to a certain level of stress beyond which they will behave in a non-linear manner. The most commonly used weldable structural materials in use are the carbon or carbon-manganese steels and these are the materials generally being considered in this book unless it is stated otherwise. This is a very large family of steels available in a multiplicity of compositions, mechanical properties, grades, qualities, etc, but in some industries these steels may just be called structural steel, perhaps divided into mild steel and high yield steel or, in industries which use only one type, just steel. Where they are introduced in this book other structural materials such as the higher alloy steels, austenitic and ferritic stainless steels and aluminium alloys are referred to specifically.

The magnitude of the stress set up in a wire or a bar under a load, P, is P divided by the cross section area, A, of the bar or wire, see Fig. 1.1.

Load
P

cross section area
A

`P`

1.1 Bar under axial load.

If the stress is σ

$$\sigma = \frac{P}{A}$$ [1.1]

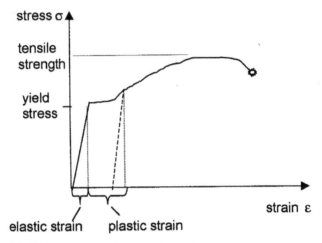

1.2 Typical stress-strain curve for steel.

In the preferred units of the international system (SI) the load will bemeasured in Newtons (N) and the cross sectional area in square millimetres (mm^2). So the stress is conventionally measured or calculated in units of N/mm^2.

If we measure the length of a bar of structural steel under an increasing tensile load and plot the stress against strain (strain = increase in length/ original length) we produce a chart such as in Fig. 1.2.

The steel shows elastic behaviour up to a certain stress but then begins to extend without the stress having to be significantly increased. This behaviour is called yielding, or plastic deformation, and the stress at which it commences is the yield stress. This yielding does not continue indefinitely and the steel begins to offer more resistance to extension until it reaches its tensile strength at which point it fractures. Once the steel has yielded it will not recover the plastic strain but will recover the elastic strain. This yielding property gives mild steel the ability to be made into products by the cold bending or forming of bar, wire, sheet and plate. The slope of the elastic part of the stress-strain curve is called the elastic modulus or Young's modulus, σ/ε, and has the units N/mm^2, typically a figure of 205 000 N/mm^2 is used for structural steels. (Note that strain being a length divided by a length has no dimensions, it is just a ratio, a number.) In the tensile test it is the load and extension which are measured and not the strain and stress. Beyond the yield point the specimen cross section is reduced by necking and the load starts to reduce although the actual stress may be increasing. The tensile strength is then an imprecise measurement because it represents the load divided by the original cross section of the test specimen which after much yielding is reduced; the calculated tensile strength is therefore an artificial figure.

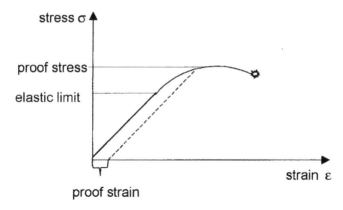

1.3 Typical stress-strain curve for austenitic steels and aluminium alloys.

In structural alloys other than the carbon steels there is no sudden departure from elastic behaviour with increasing stress; a typical stress-strain curve is as shown in Fig. 1.3. As examples, austenitic stainless steels and aluminium alloys display this type of non-linear change of strain with stress after the *elastic limit* is reached. To define a useable limit to the stress, the concept of the *proof stress* is used. This is a stress at which the material had undergone a certain permanent set, commonly 0.1% or 0.2% strain. The elastic modulus is again the ratio of stress to strain. Beyond the elastic limit the tangent to the stress-strain curve can be used to postulate a *tangent modulus* for analysing the post elastic behaviour of structures made of these materials.

These stress-strain curves, which can also be produced for compressive loading, show that a material may be used quite safely after it has been subjected to plastic or non-linear strain. Indeed this is the basis of work hardening which is used to produce materials of a higher yield or proof strength than the originally produced material. It should be noted that although the yield and proof strengths will be raised the tensile strength will not have been increased; it is important to be aware that the effect of the work hardening can be reduced or removed altogether by heating, including the heating involved in welding. Other means of enhancing the strength of engineering alloys involve metallurgical changes induced by thermal treatment or combinations of thermal and mechanical treatment and these strengthening effects can also be reversed or at least diminished by the heating from welding.

Shear stress

The stresses set up in a tensile test are stresses which are in a direction normal to the cross section, they tend to stretch the material. Shear stress resists the

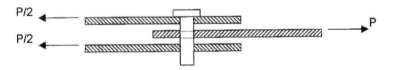

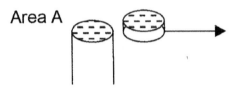

1.4 Pin in double shear.

Area A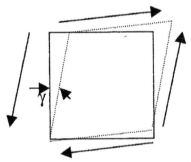

1.5 Shearing action.

tendency of a material to slide over itself as if in layers. Figure 1.4 is a simple example of shear, a pin in a tow bar. The load tries to slide the section of the pin across itself as in Fig. 1.5.

Shear stress has the same units as normal stress, N/mm^2. In this example we calculate the shear stress as the load divided by twice the cross sectional area of the pin. (Twice because it is in double shear.)

$$\tau = \frac{P}{2A} \qquad [1.2]$$

This assumes that the stress is uniformly distributed over the section which is not true but good enough for most applications.

Shear strain

We saw that normal stress does not change the shape of the object. However, shear strain does involve a change of shape. If we take a square element and apply shear to it, it changes its shape into a rhombus, see Fig. 1.6.

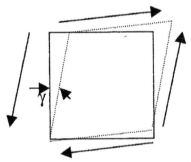

1.6 Shear strain.

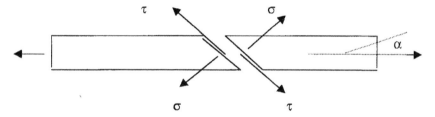

1.7 Resolved direct and shear stresses in a bar in tension.

The shear strain is the angle, γ, through which a side of the square moves and the relationship between shear strain and stress is shown in Chapter 3 in discussing the torsion of thin walled tubes.

In the simple example of a bar in tension a shear stress will be developed in any plane at an angle to the normal plane. For example, if we make an oblique cut in a bar under tension one part will tend to slide over the other as well as move away from it, see Fig. 1.7.

We can resolve the stresses across the cut into the local normal (σ) and shear (τ) stresses respectively. There is a simple way of calculating the shear and normal stresses using a diagram called Mohr's circle which plots the shear and normal stresses on a circle. The size and position of the circle on the diagram are fixed by the known stresses. In this simple case of a bar in tension we know that along the bar is a tensile stress and at right angles to this there is no (applied) stress.

For this simple case of a bar in tension, α is the angle from the longitudinal axis of the bar. When α is $0\,°$ the normal stress is at a maximum (P/A) and when α is $90\,°$, 2α is $180\,°$ and the normal stress is zero. These points then fix the diameter and centre of the Mohr's circle for this particular stress system, see Fig. 1.8. We can then see that when 2α is $90\,°$ the shear stress is at a maximum in two directions, i.e. at $\pm45\,°$ to the axis of the bar. We can measure off the maximum shear stress from the diagram. Clearly in this case it will be equal to $\sigma/2$. The stresses in the directions in which the shear stress is zero are called the principal stresses.

In a membrane such as a cylindrical pressure vessel shell normal stresses are applied in two directions (see Chapter 2). The hoop stress, σ_h, is twice the longitudinal stress, σ_l, and so the Mohr's circle is as shown in Fig. 1.9.

It can be seen that the maximum shear stress occurs at $45\,°$ to the principal planes

$$\tau_{max} = \frac{\sigma_h - \sigma_l}{2} \qquad [1.3]$$

Note that for a sphere the membrane stresses are equal in all directions and so the plot reduces to one point (equivalent to σ_l) and there is no shear stress

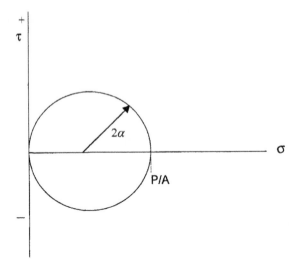

1.8 Mohr's circle for direct stress.

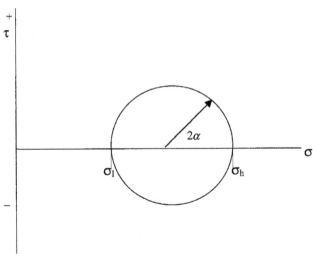

1.9 Mohr's circle for hoop and longitudinal stress in a cylinder.

anywhere. The normal stress axis can be extended to the left of the shear stress axis to allow us to plot a compressive stress. Figure 1.10 is for material with equal magnitudes of tension and compression in orthogonal directions.

The maximum shear stress is then

$$\tau_{max} = \frac{\sigma_t + \sigma_c}{2}$$

[1.4]

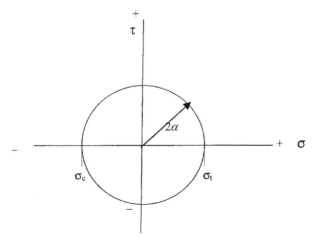

1.10 Mohr's circle for equal tension and compression.

Three dimensional stress

Up to this point in this chapter stress and strain have been considered as acting only in two dimensions. This is a reasonable assumption for thin sheet and very ductile materials but not really good enough to be able to explain some types of material behaviour, particularly in the field of fracture, where we need to be able to consider what really happens in three dimensions even though the load may be one dimensional (uni-axial). In a small round tensile test specimen the extension under load in the longitudinal axis is accompanied by contraction in the diameter. A similar contraction occurs in a piece of sheet material when axial stressing within the plane of the sheet is accompanied by contraction in the thickness. Such contractions take place up to and beyond the yield point. If the material is very thick or has no yielding behaviour this lateral contraction is not permitted to occur, which condition sets up stresses in this lateral direction. There is then instead of a *bi-axial* stress system a *tri-axial* stress system. Reflecting the stress and strain patterns the bi-axial system is referred to as having *plane stress* and the tri-axial system *plane strain*.

When a material is stretched in one direction it also contracts in the other two directions at right angles to the load. When a piece of material is acted upon by stresses σ_x and σ_y in perpendicular directions the overall elastic strains can be found by adding the strains due to the two stresses as if they were acting separately.

In the x-direction the strain is σ_x/E and the compressive strain in the y-direction is

$$\varepsilon_y = -v\sigma_x/E \qquad\qquad [1.5]$$

in which v is Poisson's ratio.

For the stress in the y-direction the strain is σ_y/E and the compressive strain in the x-direction is

$$\varepsilon_x = -v\sigma_y/E \qquad\qquad [1.6]$$

So the total strains in the two directions are

$$\varepsilon_x = \sigma_x/E - v\sigma_y/E \qquad\qquad [1.7]$$

$$\varepsilon_y = \sigma_y/E - v\sigma_x/E \qquad\qquad [1.8]$$

Multiplying by E we obtain

$$E\varepsilon_x = \sigma_x - v\sigma_y \qquad\qquad [1.9]$$

$$E\varepsilon_y = \sigma_y - v\sigma_x \qquad\qquad [1.10]$$

These equations represent what is known as a plane stress condition, which is a two dimensional system of stress which can exist in thin or very ductile materials. When we consider thick or brittle materials we must recognise that the constraint exerted on the ability of the material to contract at right angles to the applied stress will itself set up stresses. There is then a condition in which the stress system is three dimensional and the strain system is two dimensional, see Fig. 1.11. Where

$$\varepsilon_x = \sigma_x/E - v\sigma_y/E - v\sigma_z/E \quad \text{or} \quad E\varepsilon_x = \sigma_x - v(\sigma_y + \sigma_z) \qquad [1.11]$$

$$\varepsilon_y = \sigma_y/E - v\sigma_x/E - v\sigma_z/E \quad \text{or} \quad E\varepsilon_y = \sigma_y - v(\sigma_x + v\sigma_z) \qquad [1.12]$$

$$\varepsilon_z = 0 = \sigma_z/E - v\sigma_y/E - v\sigma_x/E \text{ or } E\varepsilon_z = 0 = \sigma_z - v(\sigma_y + v\sigma_x) \quad [1.13]$$

From the latter we see that

$$\sigma_z = v(\sigma_x + \sigma_y) \qquad\qquad [1.14]$$

Figure 1.12 is a Mohr's circle constructed for a three dimensional stress system. Each of the circles represents the stress on a section through one of

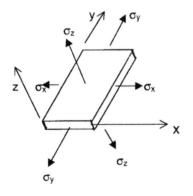

1.11 Plane strain conditions.

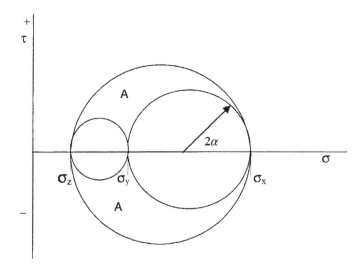

1.12 Mohr's circle for three dimensional stress.

the three axes. From this can be deduced that the largest shear stress is represented by the radius of the largest circle and is given by

$$\tau_{max} = \frac{\sigma_x - \sigma_z}{2}$$ [1.15]

It acts on the section, or plane, through the y-axis bisecting the angle through the x and z axes. Any point within the area A represents the stresses for any section inclined to the x, y and z axes.

Stress concentrations

The stress across a plain plate loaded in tension is uniform, see Fig. 1.13.

If a hole is drilled in the plate the stress close to the hole is increased by a large amount but the stress in the remainder of the plate stays much the same as before the hole was drilled, see Fig. 1.14.

For such a hole in an infinitely wide plate of an elastic material the stress

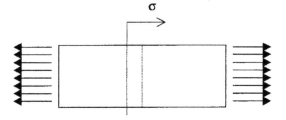

1.13 Material under uniform stress.

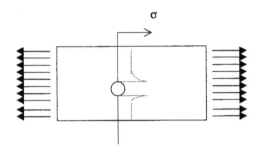

1.14 Effect of circular hole in a uniformly loaded plate.

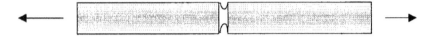

1.15 Round bar with groove.

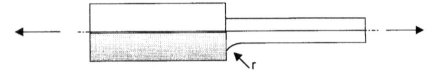

1.16 Radiused corner to reduce stress concentration.

1.17 Taper at change of thickness in plate to reduce stress concentration.

at the edge of the hole is three times the uniform stress away from the hole; that is called a stress concentration factor (SCF) of 3. A similar effect is produced in a round bar if a circumferential groove is made in it, see Fig. 1.15.

The sharper the radius of the groove the more severe is the stress concentration. A sharp change in diameter can have the same effect. Note that these two features occur in bolts: the thread is in effect a groove and the change in diameter occurs under the head where it meets the shank.

To reduce the effect we can put a radius (r) on the section change, see Fig. 1.16. In plates we may put a taper at a change in thickness for the same reason, Fig. 1.17 shows an example, but note that there is also the effect of eccentricity which may have to be allowed for.

We are most concerned about avoiding high stress concentrations where there is a fluctuating load which may cause fatigue cracking (Chapter 7) and it is most important that where possible we do not put welded joints at positions of high stress concentration. This is also a consideration in circumstances where brittle fracture has to be guarded against. There are formulae for the stress concentrations of standard shapes, and for more complicated shapes computerised finite element calculations can be used.

For materials which show a yield point or an elastic limit the position is more complicated because as the local strain increases the rate of increase of the local stress with the applied load will decrease after the yield or elastic limit is reached. In a sense this is a benefit because phenomena which result from high stress are then deferred or even avoided.

Material failure

Theories of elastic breakdown

The departure from elastic behaviour is taken as the point at which a metal specimen pulled in a tensile test starts to become permanently deformed. We may call this the yield point, or the elastic limit, the proportional limit, etc. These events are fairly easy to define in uni-axial loading as the principal stress σ_x when the other principal stresses σ_y and σ_z are zero. Fortunately this is the case, or close to it, in many simple structural parts. However there are many engineering components where more complex stress systems exist. How do we then apply the tensile test figures? We ought to know what governs elastic breakdown – what causes a material to reach its elastic limit – because it is the point at which any calculations based on elastic theory become invalid. Several theories as to when elastic breakdown occurs have been advanced such as when:

a) the *greatest principal stress* reaches a certain value. If this were true the elastic limits in pure tension and in pure shear must be the same, which they certainly are not in ductile materials.
b) the *greatest principal strain* reaches a certain value.
c) the *greatest stress difference* reaches a certain value. This is the same as saying when the greatest shear stress reaches a certain value.
d) the *total strain energy* reaches a certain value.
e) the *strain energy due to distortion* (change of shape) reaches a certain value. (This excludes change of size which is deemed to have no effect.) The basis of this has been argued as set out below.

The energy stored in a material under stress can be calculated by considering a cube of the material with sides of unit length and, so, unit volume; the work done by the stresses is the force multiplied by the distance through which it moves. The force, which on a unit face is numerically equal to the stress, starts at zero and reaches its full value at the specified strain and the work done in reaching the stress and resulting strain is ½ σε. So in a two dimensional stress system (plane stress) the strain energy, U, is given by:

$$U \; = \; \frac{1}{2} \, \sigma_x \, \varepsilon_x + \frac{1}{2} \, \sigma_y \, \varepsilon_y \qquad\qquad [1.16]$$

per unit volume of material.

Substituting for ε we obtain

$$U = \frac{1}{2} \sigma_x \left[\frac{1}{E} \left(\sigma_x - v\sigma_y \right) \right] + \frac{1}{2} \sigma_y \left[\frac{1}{E} \left(\sigma_y - v\sigma_x \right) \right] \qquad [1.17]$$

which becomes

$$U \; = \; \frac{1}{2E} \left[\sigma_x^{\,2} + \sigma_y^{\,2} - 2v\sigma_x \, \sigma_y \right] \qquad\qquad [1.18]$$

In a three dimensional stress system it can be postulated that there is a combination of fluid loading, i.e. a part of the stress equal on all faces causing change of size and the remainder which may be different on each face causing change of shape. In that case if we take the fluid stress component to be equal to the mean of the three stresses the balance of the stresses will account for change of shape. The total strain energy is

$$U \; = \; \frac{1}{2E} \left[\sigma_x^{\,2} + \sigma_y^{\,2} + \sigma_z^{\,2} - 2v \left(\sigma_x\sigma_y + \sigma_x\sigma_z + \sigma_y\sigma_z \right) \right] \quad [1.19]$$

A fluid stress which would give the same change in size would be the mean of the three stresses from which the strain energy would be

$$U^i \; = \frac{1 - 2v}{2E} \, 3 \left[\sigma_x + \sigma_y + \sigma_z \right]^2 \qquad\qquad [1.20]$$

The energy due to change in shape can then be shown to be

$$U - U^i = \frac{1 + v}{3E} \left[\left(\sigma_x - \sigma_y \right)^2 + \left(\sigma_y - \sigma_z \right)^2 + \left(\sigma_z - \sigma_x \right)^2 \right] \qquad [1.21]$$

So according to this theory the criterion is the *sum of the squares of the principal stress differences*.

Experiment shows that none of these theories covers all materials but there is evidence in favour of c) and e) for ductile materials; a), b) and d) fail to account for the fact that most materials will withstand relatively enormous fluid pressure without damage.

1.18 Ductile fracture in steel (photograph by courtesy of TWI).

Ductile fracture

In the case of the simple bar the production of a shear stress through the application of a normal stress explains the appearance of fractures in ductile materials in a tensile test. It will be found that the final fracture is often at an angle of 45° to the length of the specimen and in a specimen of rectangular cross section this sometimes results in the type of fracture as in Fig. 1.18. What is happening here is that the final failure is in shear rather than in tension and suggests that the criterion for fracture is the exceedence of a certain shear stress or strain rather than a tensile stress or strain which is commonly used in specifications as the ultimate strength of the material. In less ductile materials the final fracture will be more in the plane normal to the stress.

The fracture surface can be rough where there is considerable ductility and shear, or very smooth and crystalline in materials of low ductility such as cast iron.

Brittle fracture

Brittle fracture is a common name for the fast, unstable type of fracture which can occur in many types of material. It is best known in structural steels because depending upon the severity of residual stresses and the fracture toughness of the steel it can occur at low, or even no, applied stress. The fracture toughness of a ferritic steel can exhibit a marked reduction with a drop in temperature and brittle fracture is often, but not necessarily, associated with low ambient temperatures. The fracture surface may be smooth or coarse depending on the level of stress at which the fracture occurred, the coarsest surfaces appearing at the highest pre-fracture stresses. Figure 1.19 shows how a high proportion of the surfaces of this type of fracture present a chevron appearance in which the chevrons 'point' towards a discrete origin of the fracture which may be a crack or some material or

1.19 Brittle fracture (photograph by courtesy of TWI).

weld defect. Complex stress distributions can however distort this surface pattern and render its analysis difficult.

Fatigue cracking

As with brittle fractures, fatigue cracks tend to start at a discrete point or line representing a stress concentration. In a welded joint this may be the weld toe, the weld root or a weld defect. Under load histories other than constant amplitude the fracture surface is usually marked in bands showing the progress of the crack front from the point or line of origin. If the cracks grow over a long period of service the markings may be accentuated by coloration from corrosion or ingress of substances which the plant contains. Fatigue cracks originating from weld toes or roots tend to show steps in the earlier stages of the progression of the crack front as in Fig. 1.20 because the weld toe or root is not a straight continuous line. As a result, cracks start at various positions along the toe and then merge as they grow into one another. The final fracture will usually take the form of gross yielding or tensile failure as the maximum stress in the remaining ligament increases with the growth of the fatigue crack. Strictly speaking then there is no such thing as a structural fatigue failure: there is fatigue cracking leading to eventual fracture from the intervention of yielding or exhaustion of ductility. Failure in other senses such as leakage from a fatigue crack in a pipe can of course be considered a true fatigue failure representing unserviceability. Changes in the natural frequency of a structure due to

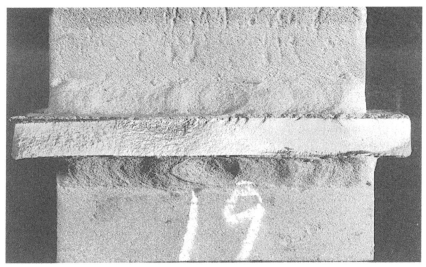

1.20 Fatigue fracture (photograph by courtesy of TWI).

the growth of a fatigue crack can lead to unserviceability or overloading due to resonance, which can lead to fracture.

Creep

Creep failure leading to cracking or fracture can occur in materials which are operated for extensive periods at high temperatures. This is usually a phenomenon which is of relevance to plant components which operate continuously under load at high temperatures such as furnaces, boilers, steam and gas turbines and their associated plant.

Summary

- There are two types of stress, normal stress and shear stress.
- Stress causes strain; under normal stress this is the ratio of the extension to the original length of the element of material being considered. Under shear stress the strain is the change in angle of the element being considered.
- Strain is proportional to stress under elastic conditions. The elastic limit is reached when the strain is no longer proportional to stress. This limit is called the yield point in common steels.
- Stress concentrations are local stress magnifications which result from geometrical features such as holes and grooves.
- Fractures in metals can be of various forms, ductile, brittle, fatigue, creep.

Problems

1.1

At a point in the web of a plate girder the vertical shear stress is 40 N/mm^2 and there is a tensile stress in the horizontal direction of 100 N/mm^2. Using Mohr's circle calculate the principal stresses and the maximum shear stress in the web.

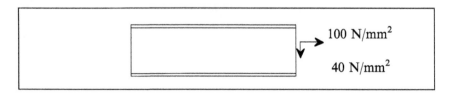

1.2

A works manager wants to build an air receiver from a length of spirally welded pipe, 600 mm diameter and 3 mm wall thickness. If the gauge pressure is 10 bar what is the shear stress along the weld? Assume that at all points on its length the weld makes an angle of 45° with the generator.

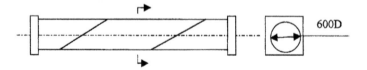

Stresses in some common types of structures

Frames

From the structural engineering aspect a frame is an assembly of bars arranged so that they can support a load. The simplest type of frame is one where the joints are effectively pins so that the loads in the members are entirely axial. Figure 2.1 shows a simple frame which can be seen in many commonplace applications.

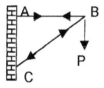

2.1 Simple frame.

2.2 Resolving forces at B in the vertical direction.

If the length AB = length AC the angles at C and B are 45 °. We can find the load in each member by resolution of the load at B. In the vertical direction as in Fig 2.2, if the force in BC is F_{BC} then

$$P = F_{BC}/\sqrt{2} \qquad\qquad [2.1]$$

or

$$F_{BC} = P\sqrt{2} \qquad\qquad [2.2]$$

In the horizontal direction as in Fig. 2.3

$$F_{AB} = F_{BC}/\sqrt{2} \qquad\qquad [2.3]$$

2.3 Resolving forces at B in the horizontal direction.

which we know from the vertical sum $= P$.

We could reach the same result by taking moments about C

$$P \times (AB) = F_{AB} \times (AC) \qquad [2.4]$$

and since $AB = AC$

$$P = F_{AB} \qquad [2.5]$$

and

$$F_{BC} = P/\sqrt{2} + F_{AB}/\sqrt{2} \qquad [2.6]$$

$$= 2P/\sqrt{2} = P\sqrt{2} \qquad [2.7]$$

To complete the frame we recognise that the wall between A and C is the third member.

We can build up larger frames from this basic triangle. Figure 2.4 will be recognised as a simple frame often used in lightweight bridges. The members shown as dotted lines do not contribute to the load carrying capability of the frame and can be dispensed with unless they are required as safety barriers to stop people falling off the bridge.

Although rendering the calculations very simple, pin joints are not satisfactory in practice for a permanent structure. For such a structure the chords may be continuous and the bracing bolted or welded to the chords. Then, when the bridge deflects under load, bending will be set up in the members and is known as secondary bending. This may have to be taken into account when designing the joints and the member sections and requires a more complicated calculation procedure which is usually computerised.

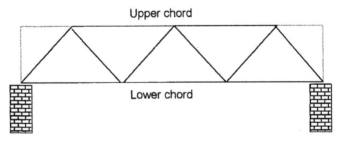

2.4 Simple bridge structure.

Pressurised shells

Besides the structures already mentioned, probably the other most common form of welded fabrication is the pressure vessel whether in the form of a discrete container or as pipework and pipelines. Examples of pressurised containers range from the common aerosol can, pneumatic reservoirs for lorry braking systems, oil storage tanks, spherical gas containers, submarine hulls, aircraft fuselages, chemical process reaction vessels, steam boilers, heat exchangers, water towers and silos.

Pipework and pipelines appear in many industries and we might include with them penstocks for hydropower schemes, although their complexity might sometimes suggest they be treated more as pressure vessels. However for the purposes of this book these distinctions are unimportant. The stresses in the body of a simple pressurised cylinder, see Fig. 2.5, are easily calculated. Taking a section across a diameter the only reaction against the pressure is the hoop load in the shell, P, see Fig. 2.6.

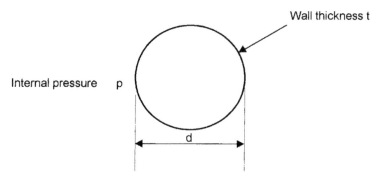

2.5 Cylinder under pressure.

2.6 Hoop stress in cylinder under pressure.

By simple statics, resolving the pressure loads across a diameter

$$pd = 2P \qquad [2.8]$$

or

$$P = pd/2 \qquad [2.9]$$

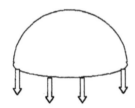

2.7 Sphere under internal pressure.

The *hoop stress* is then

$$\sigma_\lambda = \frac{pd}{2t} \qquad\qquad [2.10]$$

and for a pressurised sphere, see Fig. 2.7, taking a section across a diameter the load across the diameter is the pressure p on the projected area of the sphere and is taken by the wall of the sphere around the diameter.

The pressure load is

$$p\pi d^2/4 \qquad\qquad [2.11]$$

and is reacted by the wall of length πd and thickness t.

The stress in the wall is then

$$\sigma = \frac{p\pi d^2/4}{\pi dt} = \frac{pd}{4t} \qquad\qquad [2.12]$$

which, as we can see, is half of the hoop stress in a cylinder. This difference has a practical implication because one of the easiest ways of closing the end of a cylinder is to put a hemisphere on it. However, we can see that when such a vessel is pressurised the hoop strain in the cylinder is twice the circumferential strain in the end, or head, which will give a mismatch. In an exaggerated form this will appear as in Fig. 2.8.

This will set up local bending stresses in the shell and so transition shapes are used to reduce these local stresses. This transition point is usually where

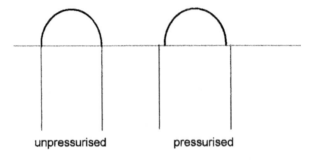

unpressurised   pressurised

2.8 Differential strain between cylindrical body and spherical head.

welded joints are required and so these may be subject to some of the most severe stress conditions in the vessel. This is particularly significant if the loads are repeated and fatigue cracking becomes a design criterion.

We have seen that the *longitudinal stress* in a cylinder is the same as the hoop stress in a sphere, i.e. half of the hoop stress in the cylinder. This is in some ways a benefit in pipework and particularly in pipeline construction. Any site welding on line pipe is in the circumferential welds which react the longitudinal stress which is half the hoop stress and may be even less if the pipe is supported against the longitudinal thrust of the pressure of the contents. This may be of significance if an engineering critical assessment is being made in respect of weld defect acceptance levels and may allow the defect acceptance levels in the circumferential welds to be less demanding than in the longitudinal welds.

Pressure vessels have pipes connected to them via *nozzles* to allow the introduction and extraction of the process materials; pipes and pipelines have *branches* where they divide or collect. These features create complicated local stress fields because of the competing strains, as we have seen with the simple case of the hemispherical end to a cylinder. As a result, these details tend to be designed by reference to standards or codes of practice for which the mathematics have already been worked out and agreed. Unusual shapes or details may still have to be analysed by finite element methods.

The nature of many of the details at nozzles and branches creates high stress, or strain, concentrations requiring parent and weld metal properties and weld qualities able to sustain these high local strains. For carbon-manganese steel pressure vessels an allowable stress for the straightforward sections is set at about two-thirds of yield. Nozzles and other details are then designed so that the stress concentration effect will keep the material working within its capability. A feature of many processes employing vessels and pipes is that they take place at temperatures well above or well below ambient which, in this context, is the very small temperature range of between, say, 0 °C and 100 °C. This difference from ambient, or changes occurring during a process, introduces thermal expansion and contraction as a source of load and deflection and also introduces the need for materials and weld metals which will offer the necessary strength and integrity at these temperatures. For lower than ambient temperatures we are looking at fine grain notch tough steels, chrome/nickel ferritic steels, austenitic steels, and aluminium and its alloys. At higher than ambient we are looking variously at the elevated temperature carbon-manganese steels, the chrome-moly steels, a range of nickel alloys and other more esoteric materials.

Tubular structures

The marriage of structural and pressure vessel material specifications and fabrication technology was forced in the 1970s for the design of the first generation of deep water platforms for the northern North Sea oil and gas production. These platforms followed traditional designs from shallower waters using round steel tubes to construct legs and braces. The legs were big tubes and the braces smaller ones. The need for the introduction of pressure vessel technology arose due to the following features:

- large structures, therefore thick sections
- local high stress concentrations
- temperatures at the lower boundary of our 'ambient' range
- structures supporting process plant with all the potential hazards of an oil refinery with people living literally on top of it and unable to escape quickly in the event of an incident
- limitations of in-service underwater inspection and repair techniques
- plant representing a large capital investment by the operators requiring continuing revenue generation from uninterrupted oil production.

So what are the similarities between the platforms and the pressure vessel? They can be summarised as:

- material thickness and generic type
- joint details, i.e. for vessel read 'chord' or 'node can' and for nozzle read 'brace'.

And the differences from the pressure vessel can be summarised as follows:

- there is no hole in the chord at the brace
- large axial and bending loads in braces
- no significant internal pressure loads
- complex load history
- little scope for intensive routine inspection and maintenance
- no scope for proof loading.

The loads from gravity, wave action and current are used to develop the structural design using formulae published in documents such as BS EN ISO 19902 *Petroleum and natural gas industries. Fixed steel offshore structures.* Local secondary stresses in the chord and the brace are caused by the incompatibility of the local strains set up by the out of plane loads acting on the surface of the chord, Fig. 2.9.

The overall sizing of the joint is based on published formulae and, if thought necessary for fatigue life calculations, the peak stresses in these 'nodal' joints are derived either from published parametric formulae or by finite element analysis.

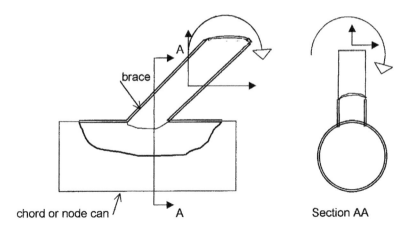

2.9 Nomenclature of tubular joint.

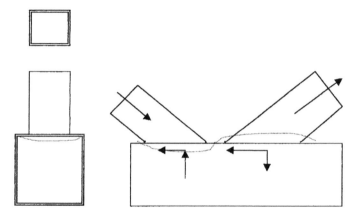

2.10 Deformations of members in joint between rectangular hollow sections.

Many of the tubular structures built on land are made from circular or rectangular hollow sections. These tubes are generally much smaller than those used in offshore platforms but there are particular considerations in the joining of square or circular braces to the face of the rectangular or square section chord.

The brace pushes or pulls on the face of the chord causing high local strains, particularly in between the two braces, see Fig. 2.10. One solution to this – which is not as easy to fabricate – is to overlap the braces, as in Fig. 2.11, so that they react between themselves a proportion of the component of the loads at right angles to the chord which depends upon the degree of overlap.

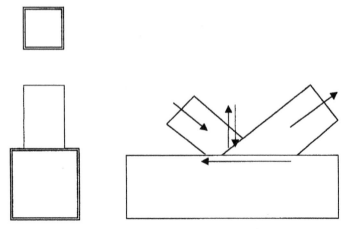

2.11 Reactions in overlap joints between rectangular hollow sections.

Effect of deviations from design shape

Chapter 5 shows how fabrications can rarely be made to the exact shape required by the drawing and that some level of dimensional tolerance has to be set. Dimensional tolerances will depend on the function of the component and the capability of the fabrication processes and may have to be a compromise between the desirable and the possible.

Deviations from design shape may be manifested as linear or angular misalignment or departure from design length or cross section. Are these deviations important? The answer to this depends on the nature and function of the fabrication. Matching up with adjacent assemblies to which it is connected is clearly a fundamental reason for getting things the right size and shape. Interchangeability is also another sound reason; when a part has to be replaced with another during service we need to be able to just slot it into place. The performance of the product may depend on shape, as in the case of a ship's hull. However, we also need to consider the effect of deviations from design shape on the strength of the product. As an example, think about a column holding up a building. It is designed to be straight, within practical limits, and the section supports the weight of the building and resists some of the end moments from the floor joists. If the column is not straight, the eccentricity of the load will cause bending in the column which may result in the allowable or limit compressive stress being exceeded. A slender column may even become unstable and buckle.

In the exaggerated situation in Fig. 2.12 the eccentricity set up by the curvature of the column is e and so the bending moment set up in the column is Pe. Instability, or buckling, will occur if the deflection caused by the moment causes the moment to continually increase. This may be in the elastic or plastic range of material behaviour. Plate and box girders and box

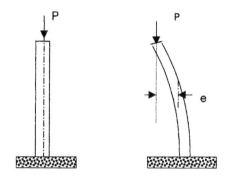

2.12 Eccentricity in column.

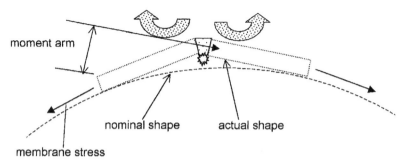

2.13 Roof topping as a source of local bending.

columns may also suffer from instability not only due to deviations from shape but as a result of the residual stresses. The design rules for these as used in bridges take into account the effect of residual stress. Since post weld heat treatment of large box girders is impracticable we have to live with this effect. This leaves us then with a need to control the dimensions of the fabrication and in particular the flatness of the panels.

Mismatch between plates gives rise to eccentricities which can then set up local, secondary bending moments which may not have been taken into account in the design. Generally speaking these do not matter if the joint is in static tension, however they may become significant if there is a fluctuating load which can then set up fatigue cracking. This is a particular concern in round pipes and vessels if '*roof topping*' occurs in the plates at the longitudinal welds when the plate rolling or other bending method does not curve the plate right up to its end. This is described further in Chapter 5 but from here the concern is that the resultant shape induces bending stresses, see Fig. 2.13, in the nominally circular shell which is designed for membrane stresses only. Under fluctuating loads these bending stresses can induce fatigue cracks at the weld toe or at the weld root of one sided welds. They may also give rise to other forms of failure such as brittle fracture. Excessive ovality and other variations in radius can of course introduce bending into a cylindrical shell of a magnitude related to the degree of deviation from circularity.

Summary

- For convenience of calculation we can think of structures as frames, beams or plates.
- Members in frames are assumed to carry axial load only.
- The stress in the walls of pressurised spheres and cylinders can be calculated by a simple equation. In a cylindrical vessel with spherical heads (ends) the strains are not compatible and designs have to recognise this.
- The joints between tubes forming braced structures must be designed to allow for the local stress concentrations induced by the incompatibility of strains at the joint.
- Dimensional tolerances must be placed on structural members to control local stress concentrations or prevent load magnification.

Problems

2.1

You are travelling in an aircraft in the front row of seats; the fuselage cross section appears to be circular. You estimate that the diameter of the fuselage in your area is 6 m. The typical pressure difference between the cabin and the outside air in a civil aircraft is 0.6 bar. If the allowable tensile stress in the fuselage skin is 150 N/mm² how thick is it (ignoring for the sake of this example the stress concentration effect of windows, doors, rivets etc)?

2.2

Calculate the load in this bicycle cross bar assuming that the joints are pinned. The cyclist weights 75 kg.

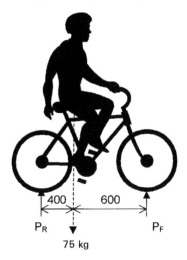

Elementary theories of bending and torsion

Bending

We saw in Chapter 1 how the stress in a member in tension is calculated; however, few members in any structure are loaded simply in tension. Most structures consist of elements which can be reduced in their effect to beams or plates. Probably the simplest form of structural element is the beam, which we see in the form of the builder's scaffold plank and the floor joists in a house. The beam, see Fig. 3.1, is supported at each end and carries the load at any point or all along its length.

3.1 Simple beam.

Under the weight of a person (the load) the beam flexes, as in Fig. 3.2.

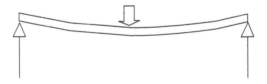

3.2 Deflected beam.

We know that this flexibility is different for different cross sections. For example, if we use the plank as in Fig. 3.3, bending it about its *major axis*, it will not flex as much as if we use it as in Fig. 3.4, bending it about its *minor axis*. These terms are explained later in this chapter.

So let us look at what is going on in the beam (our plank). When the beam flexes it is loaded in bending. Perhaps an easier way of modelling this

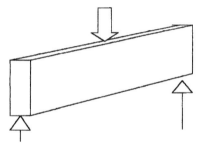

3.3 Beam in bending about major axis.

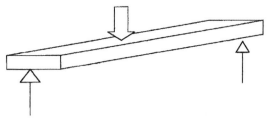

3.4 Beam in bending about minor axis.

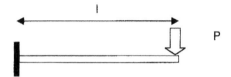

3.5 Springboard as an example of a cantilever.

3.6 Root reaction on cantilever.

3.7 Root moment on cantilever.

is to think of a bar clamped at one end as a cantilever, like the springboard in Fig. 3.5.

The fixed end of the beam does not move up or down or rotate so there must be a set of loads or reactions keeping that end where it is. Stopping the beam going up or down is the vertical reaction V in Fig. 3.6.

Stopping the beam rotating is the moment M in Fig. 3.7.

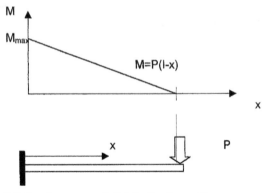

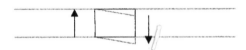

3.8 Bending moment distribution in cantilever.

Now the root bending moment $M = Pl$ and the graph of the bending moment along the beam is shown in Fig. 3.8.

The vertical reaction V creates a shear force in the beam which tends to deflect the beam as in Fig. 3.9.

3.9 Shear deflection in beam.

In practice this deflection is small enough not to be noticed in most beams but the moment has a more obvious effect; in our example shown in Fig. 3.10 it bends the beam by stretching the top 'fibres' and compressing the bottom 'fibres'.

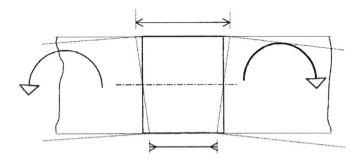

3.10 Bending deflection in beam.

At the centre plane of the beam there is no stretch or compression. This is called the *neutral surface* and the line where it intersects the cross section is called the *neutral axis* (NA), see Fig. 3.11.

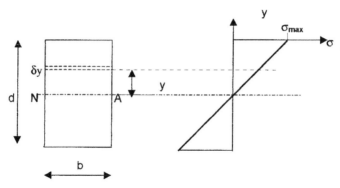

3.11 Stress action across section in bending.

We have seen in Chapter 1 how when an elastic material is stretched (or compressed) a stress proportional to that stretch is produced in the material. The stress in the beam (our springboard again) at any distance from the centre (y) through its depth (d), which we will call σ, can then be calculated on the assumption that the amount of stretch in each layer of the beam is proportional to its distance from the neutral axis. This is shown in Fig. 3.11 where at any distance from the neutral axis, y, the stress is

$$\sigma_y = \sigma_{max} \frac{y}{d/2} \qquad [3.1]$$

We can calculate the moment resisting the bending of the beam by pretending that it comes from the stress in narrow strips δy deep, which each have area $\delta y \times b$ and are y from the neutral axis. The moment from one strip is then $\delta y b \times \sigma \times y$ and by adding the strip in compression on the opposite side of the neutral axis the moment is then

$$M = 2\delta y\, b\, y\, \sigma = 2\delta y\, b\, y\sigma_{max} \frac{2y}{d} = 4\delta y\, b\, y^2 \frac{\sigma_{max}}{d} \qquad [3.2]$$

Integrating over the whole depth of the section the total moment is then

$$M = \int_0^\sigma 4b\sigma_{max}\, y^2/d\, \mathrm{d}y \qquad [3.3]$$

Now we know that the moment of inertia of the beam cross section, I, is

$$\int_0^d 2y^2\, b\, \mathrm{d}y \qquad [3.4]$$

and so

$$M = 2\sigma_{max}\, I/d \qquad [3.5]$$

The expression

$$I/d/2 = Z \qquad [3.6]$$

is called the *section modulus*.

And so

$$M = \sigma_{max} Z \qquad [3.7]$$

This gives the maximum stress in the beam in two senses; it is the stress at the extreme fibre of the cross section and also the stress where the bending moment is highest, i.e. at the fixed end of the beam. We can extend this type of calculation to other section shapes such as rolled and fabricated steel beams and columns by calculating the values of I for the component parts of the section and adding them up, as in the example later in this chapter.

We have seen in Chapter 1 that steels will stretch elastically up to their yield point and then stretch plastically. If we assume that at any point at a distance x along the beam as shown in Fig. 3.12 it is bent in a circular arc, we can derive a useful equation as explained below.

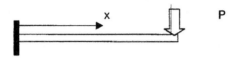

3.12 x-axis of beam.

For small amounts of bending over a very small length of the beam there will be a rotation giving a local radius of the beam, R, see Fig. 3.13.

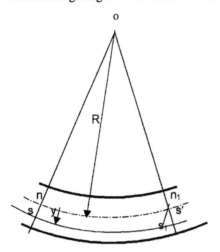

3.13 Elemental length of beam in bending.

The strain in the fibres at a distance y from the neutral axis ε, is the extension divided by the original length, i.e. $s_1 \, s' / n \, n_1$.

From similarity of the triangles $n \, o \, n_1$ and $s_1 \, n_1 \, s'$

$$\varepsilon = \sigma/E = y/R \qquad [3.8]$$

also

$$\sigma_{max} = My/I \qquad\qquad [3.9]$$

We can rearrange these to give the relationships

$$\frac{\sigma}{y} = \frac{M}{I} = \frac{E}{R} \qquad\qquad [3.10]$$

This is the basic equation for all simple beam calculations.

As well as the stress in the material we are interested in the deflection of the beam. We saw above that under a bending moment the mean radius can be shown to be given by

$$\frac{1}{R} = \frac{M}{EI} \qquad\qquad [3.11]$$

In Fig. 3.14 o is the centre of curvature of the beam with a radius of curvature r.

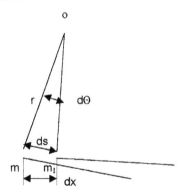

3.14 Deflection of element of beam in bending.

Then

$$ds = rd\Theta \qquad\qquad [3.12]$$

or

$$\frac{1}{r} = \left| \frac{d\Theta}{ds} \right| \qquad\qquad [3.13]$$

The bars indicate that this is the numerical value of the curvature. We have taken the applied bending moment to be positive when the beam is bent downwards.

In practical engineering deflections are very small so we can assume that

$$ds \approx dx \text{ and } \Theta \approx \tan\Theta = dy/dx$$

Substituting these values for ds and Θ we obtain

$$\frac{1}{r} = -\frac{d^2 y}{dx^2} \qquad\qquad [3.14]$$

and so

$$EI \frac{d^2 y}{dx^2} = M \qquad\qquad [3.15]$$

We saw that for the 'springboard'

$$M = P(l - x) \qquad\qquad [3.16]$$

So

$$EI \frac{d^2 y}{dx^2} = P(l - x) \qquad\qquad [3.17]$$

By multiplying both sides by dx and integrating we obtain

$$EI \frac{dy}{dx} = P(lx - x^2/2) + C \qquad\qquad [3.18]$$

At the fixed end of the beam there is no slope, i.e.

when $x = 0$

$$\frac{dy}{dx} = 0 \qquad\qquad [3.19]$$

And so

$$C = 0 \qquad\qquad [3.20]$$

By integrating again

$$EIy = P(l\,x^2/2 - x^3/6) + D \qquad\qquad [3.21]$$

Again

when $x=0$, $y=0$ and so $D=0$

At the free end $x = 1$ and $y = y_{max}$

$$y_{max} = \frac{P}{EI}(l^3/2 - l^3/6) = \frac{Pl^3}{3EI} \qquad\qquad [3.22]$$

The same procedure can be followed for other types of support and loading. However, as a short cut we can look at our original plank as two fixed ended beams, see Fig. 3.15.

By symmetry the centre will act as if it were fixed against rotation and so the deflection is still

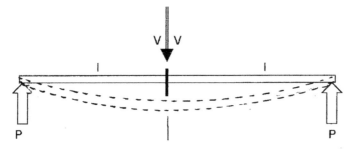

3.15 Simply supported beam deflected under central load.

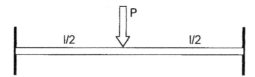

3.16 Nomenclature of beam loads and dimensions.

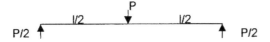

3.17 Fixed ended beam under central load.

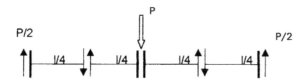

3.18 Fixed ended beam broken down into four cantilevers.

$$y_{max} = \frac{Pl^3}{3EI}$$

as equation [3.22]. If we now use the nomenclature P for a single load and l for the span, see Fig. 3.16, the central deflection in these terms is

$$\frac{P/2 \times l^3/8}{3EI} = \frac{Pl^3}{48EI} \qquad [3.23]$$

If our beam is built in at its ends, see Fig. 3.17, so that its ends cannot rotate we can still use the springboard element to calculate its deflection. This is equivalent to four 'springboards' as in Fig. 3.18. The deflection at the centre is then twice the deflection of each 'springboard'

$$2\frac{P/2 \times l^3/64}{3EI} = \frac{Pl^3}{192}$$

[3.24]

Calculating *I* for other sections

If we look at the stress distribution throughout the cross section of the rectangular section it is clear that only the outer fibres are working to their full capability. This gives rise to the idea that for an efficient beam as much material as possible should be placed at the furthest distance from the neutral plane. If all the material in each half of the section were displaced to the extreme fibre, i.e. at a distance of $d/2$ from the neutral plane, then

$$I = 2.\frac{A}{2}.\left(\frac{d}{2}\right)^2 = \frac{Ad^2}{4},$$

[3.25]

$$Z = \frac{Ad}{2}$$

[3.26]

This is a limit which may be approached in practice by the use of I beam and column sections. We will now consider an example of how the properties of an I section can be calculated; the section will be assumed to be made up of three rectangular elements, two flanges and one web, see Fig. 3.19. In practice rolled sections will have fillets at the intersections as will welded plate girders but we will ignore them both for clarity here and because they will add little to the final answer.

We will use the following dimensions in millimetres:

B: 300
D: 600
t_f: 30
t_w: 20

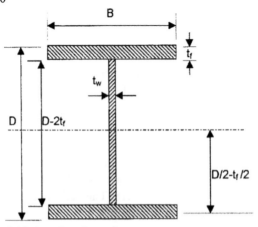

3.19 *I* section dimensions.

The distance of the centroid of an item from the neutral axis is $+y$ above and $-y$ below.

The I value for a rectangle of height d and width b is

$$I = \frac{bd^3}{12}$$ [3.27]

The total I value for the section is the sum of the item areas each multiplied by the square of its distance from the section neutral axis plus the I value for each item about its own neutral axis. This is used in the following calculation for deriving the I values in the last column for each item:

Item	Area	Dimensions, mm	A	y	y^2	$Ay/10^3$	$Ay^2/10^6$	$I/10^6$
Flange	$b.t_f$	300×30	9000	285	81225	2565	731.025	0.675
Web	$(d-2t_f)t_w$	540×20	10800	0	0	0	0	262.000
Flange	$b.t_f$	300×30	9000	-285	81225	-2565	731.025	0.675
Total	-	-	28800	-	-	0	1462.050	263.350

Total $I = Ay^2 + I = 1725.4 \times 10^6$ mm^4

Compare the I value of a rectangular section of the same cross section area and depth D. For the same area the width will be $28800/600 = 48$ mm.

Then

$$I = \frac{48 \times 600^3}{12} = 864 \times 10^6 \text{ mm}^4$$

Which is about half that of the I section with the same volume of material.

Shear stresses in beams

So far we have looked at the bending stresses but in the examples above the bending is caused by a transverse load which will also set up shear stresses in the beam. As a very crude method of calculating the shear stress in the rectangular section beam we can just divide the shear load by the cross section area

$$\tau = \frac{P}{bd}$$ [3.28]

and for our I beam we can just assume that the shear is reacted by the web, extended into the flanges, see Fig. 3.20.

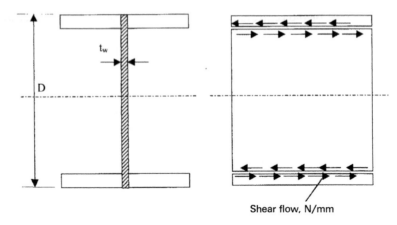

Shear flow, N/mm

3.20 Gross web for shear calculation.

And so

$$\tau = \frac{P}{Dt_w}$$ [3.29]

In a fabricated girder the shear transfer between the web and the flange is via the web to flange weld and if this is a twin fillet weld we need to know what shear load has to be carried across the weld so that we can specify its size. We could use the crude method above. The shear flow, q, is then

$$\tau t_w = \frac{Pt_w}{Dt_w} = \frac{P}{D} \text{ N/mm}$$ [3.30]

So the fillet weld has to be able to carry a shear flow of this magnitude. There is a more accurate method of calculating the web to flange shear flow which can take into account the effect on the shear flow of the relative proportions of the flanges and the web. If we assume that a beam is made of two layers which can slide over each other, if a transverse load is applied to it the layers will slide, see Fig. 3.21, and the beam will not be able to support the load as effectively as if it were made of one single solid bar. This shows that there is a shear stress in the solid bar which is maintaining the cross sections in their original form.

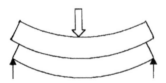

3.21 Potential shearing of a beam.

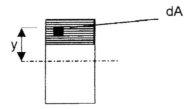

3.22 Excluded area of section in shear.

To find out about this shear stress let us take a section, which we will call the *excluded area*, shown shaded in Fig. 3.22, and find out what shear stress keeps it from sliding over the remainder of the section.

The normal force acting on an elemental area dA is

$$\sigma dA = \frac{My}{I}\, dA \qquad [3.31]$$

The sum of all these forces acting over the excluded area will be

$$\int_{y_1}^{d/2} \frac{My}{I}\, dA = \frac{M}{I} \int_{y_1}^{d/2} y dA = \frac{MA\bar{y}}{I} \qquad [3.32]$$

Where $A\bar{y}$ is the first moment of the excluded area about the neutral axis of the beam. The change in this resultant force per unit length of the beam is

$$\frac{d}{dx}\left(\frac{MA\,\bar{y}}{I}\right) = \frac{PA\,\bar{y}}{I} \qquad [3.33]$$

The shear stress along the boundary of the excluded section is then

$$\tau = \frac{PA\,\bar{y}}{Ib} \qquad [3.34]$$

As an example, put a load of 23 tonnes in the centre of a 20 m span of the beam section which we used above, see Fig. 3.23.

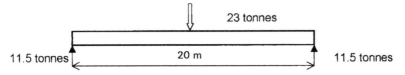

3.23 Beam supporting load.

The maximum bending moment is 23 × 20/4 tonne metres, i.e

$$115 \times 10^7 \text{ N mm}.$$

Checking the bending stress

$$\sigma_{max} = \frac{115 \times 10^7 \times 300}{1725 \times 10^6} = 200 \text{ N/mm}^2$$

the shear is 11.5 tonnes and the shear flow is then

$$\frac{11.5 \times 10^4 \times 2565 \times 10^3}{1725 \times 10^6} = 171 \text{ N/mm}$$

and the shear stress in the web along the web to flange junction will be

$$171/20 \approx 8.5 \text{ N/mm}^2$$

If the web were attached to the flange by two 10 mm leg length fillet welds the weld throat area is $20/\sqrt{2} \approx 14$ mm. The weld shear stress would then be

$$171/14 \approx 12 \text{ N/mm}^2$$

Using the crude method above the shear flow would be $11.5 \times 10^4/600 \approx 192$ N/mm which is some 12% different. In this type of beam section the shear in the web is actually greatest at the neutral axis and if the web plate is, for example, to be made from two plates joined along their length by a partial penetration weld then it would be prudent to calculate the actual shear flow along the weld.

Torsion

Besides axial load and bending a structural member may be subjected to torsion, in effect a moment acting about the longitudinal axis of a member such as the tube in Fig. 3.24.

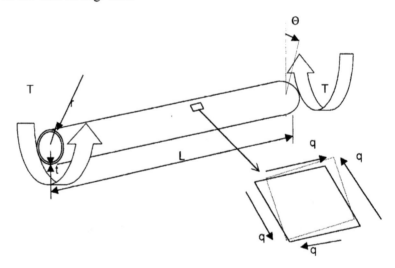

3.24 Thin walled tube in torsion.

A shear stress, $\tau\,(\mathrm{N/mm^2})$ is developed in the circumferential direction and a complementary shear stress along the generator; q, the *shear flow*, is $\tau t\,(\mathrm{N/mm})$. For a thin walled circular tube the torque is reacted by the shear stress over the whole circumference acting about the longitudinal axis of the section at a radius r.

$$T = \tau t.r.2\pi r = \tau t.2\pi r^2 \qquad\qquad [3.35]$$

and since $\pi r^2 = A$, the area inside the tube wall

$$T = 2\tau t.A \qquad\qquad [3.36]$$

or

$$\tau = \frac{T}{2At} \qquad\qquad [3.37]$$

and so

$$q = \frac{T}{2A} \qquad\qquad [3.38]$$

It can be shown that the rotation, Θ (radians), between the ends of the tube is given by

$$\frac{T}{J} = \frac{G\Theta}{L} = \frac{\tau}{r} \qquad\qquad [3.39]$$

where J is the polar moment of inertia and G is the shear modulus. It can be seen that this expression is analogous to that for the bending of beams. The complementary, or longitudinal, shear can be experienced by rolling up a sheet of paper into a tube and then twisting it. Note that the tube has no torsional strength unless the free end of the paper is glued down. Without such a joint the tube is free to *warp* and the equations here are not valid; this accounts for the weakness in torsion of open sections such as channels. Warping also occurs in non-circular hollow sections and has to be taken into account when calculating the shear stress in such sections.

Vibration

For an object to vibrate it must be acted upon by a restoring force such as might be provided by a spring, which might be part of the object itself. If the object is displaced from its normal position and then released it will, if its equilibrium is stable, oscillate about its position of equilibrium. This oscillation is termed *free oscillation* but under the influence of damping, which is always present to some extent, the free oscillations will sooner or later subside into insignificance. As an example take our original beam, or

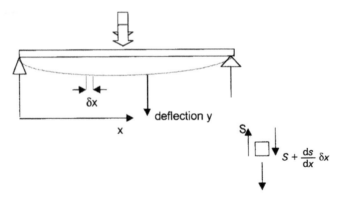

3.25 Vibrating beam.

plank, see Fig. 3.25. If the user jumps off it the board will oscillate at its own natural frequency and the oscillations will quickly die away as a result of damping from the air and the mountings. If the oscillation is to be maintained it is necessary to provide a periodic force which will create a *forced oscillation*. If we hold onto the board we can move it up and down thus producing a forced oscillation, but nothing else will happen to it. However, if we stand on it and jump up and down at various rates we may find one rate of jumping which causes the board to deflect in time with our jumps and we will find that the deflection of the board increases quite markedly. We have found a natural frequency of the board and our forcing frequency is the same.

We can show this mathematically: consider the motion of the element of length δx. If S is the shear force, M the bending moment and m the mass per unit length, the net force acting on it is

$$S_{net} = S + \frac{ds}{dx} \delta x - S = \frac{ds}{dx} \delta x \qquad [3.40]$$

and its acceleration is

$$f = \frac{d^2 y}{dt^2} \qquad [3.41]$$

Its mass is $m\delta x$.

And so, force being mass times acceleration

$$\frac{dS}{dx} = m \frac{d^2 y}{dt^2} \qquad [3.42]$$

and since

$$M = -EI \frac{d^2y}{dx^2} \tag{3.43}$$

and

$$S = \frac{dM}{dx} \text{ and so } \frac{dS}{dx} = \frac{d^2M}{dx^2} \tag{3.44}$$

then

$$-EI \frac{d^4y}{dx^4} = m \frac{d^2y}{dt^2} \tag{3.45}$$

is the equation of motion.

This would be satisfied with a solution of the form

$$y = A \sin \frac{p\pi x}{l} \sin 2\pi nt \tag{3.46}$$

for this expression gives $y = 0$ and $M = 0$ when $x = 0$ and $x = 1$ which is the case with the beam

$$M = -EI \frac{d^2y}{dx^2} = 0 \tag{3.47}$$

Putting the expression for y into the equation of motion we can obtain

$$EI \frac{p^4\pi^4}{l^4} = 4\pi^2 n^2 m \tag{3.48}$$

If we take the square root and re-arrange we have

$$2\pi n = \frac{p^2\pi^2}{l^2} \sqrt{\frac{EI}{m}} \tag{3.49}$$

By putting $p = 1, 2, 3, 4$ etc we find a range of natural frequencies n_1, n_2, n_3, n_4 etc and by putting them in terms of the fundamental frequency which is given by

$$2\pi n_1 = \frac{\pi^2}{l^2} \sqrt{\frac{EI}{m}} \tag{3.50}$$

the range is $n_1, 4n_1, 9n_1, 16n_1$ etc and the modes of vibration have wave lengths of $2l, 4l, 6l$ and $8l$.

Summary

- The stress in a beam due to bending, and its deflection, can be calculated from the load and the moment of inertia of the cross section.
- The shear stress in a beam on any plane can be calculated using the moment of inertia of the cross section and the 'excluded area'.
- The natural frequency of vibration of a beam can be calculated using the same quantities.

Basis of design of welded structures

Allowable stress design

In the earliest days of iron or steel use in structures the design was directed at the avoidance of failure in the sense of the fracture of components, be they girders in bridges or the shells of boilers. It was accepted that the consistency of materials was variable and the knowledge of the loads and the distribution of stress was imprecise. It was therefore the practice to design so that the maximum calculated stress did not exceed a certain fraction of the tensile strength – typically a quarter. An analogous approach is in use today except that instead of a fraction of tensile strength, a fraction of the yield stress is used for structural steels and a fraction of a proof stress for structural materials such as stainless steels and aluminium alloys which do not exhibit a clear yielding phenomenon. This approach recognises the closer control of material properties which has come to exist and the significance of yield or proof stress as a more realistic and controllable limit to the useful working condition of the material than tensile strength, as well as recognising the greater reliability of load prediction and stress calculations.

The examples with the 'springboard' in Chapter 3 describe a situation where initially the stresses are elastic; to whatever deflected shape the beam goes it will return to its original straightness. The designer will choose a cross section which, under the load, will not be stressed above the yield point of the grade of steel used. In reality the control of the load, the accuracy of the calculations and the tolerances on materials and fabrication all introduce elements of uncertainty in the prediction of the stress actually developed. The designer cannot therefore safely design right up to the yield stress. Some margin must be allowed to ensure that under actual conditions the stresses remain below yield. A common way of dealing with this is to design to a considerably lower stress than yield, typically a stress of 2/3 of the yield stress. For a mild steel this will be in the region of 165 N/mm^2 in tension. This figure has no rational basis and is based on judgement and experience. This design method or philosophy is called the *allowable stress* or the *elastic design* method. It is relatively simple and is still used in many industries.

Although the allowable stress method appears to be based on the stress nowhere exceeding two-thirds of the yield stress, there are areas in a real structure where this stress is exceeded due to stress concentrations (see Chapter 1). This dismissal of what may be thought to be a significant departure from the safety inherent in the limitation of the maximum stress is justified on the basis that, provided the material has a degree of ductility, what results in failure is strain and not stress. If the stress concentration is local, for example a bolt hole, the local strain is limited by the overall elastic strain in the surrounding structure. In structural steels the locally magnified strain is well within the plastic range of behaviour and no damage occurs. Similarly for other materials there is a non-linear part of the stress/strain behaviour which can accept the enhanced local strain.

A key condition of the use of the allowable stress method is that the design is competently executed so that there are no areas of stress concentration unconstrained by surrounding elastic strains which could lead to progressive failure under steady load or to incremental collapse under alternating loads. In the power generation and process industries where pressurised plant such as chemical process reaction vessels, steam boilers, turbines and heat exchangers are used, both economy in design time and safety are served by the use of standardised types of detail which have been shown to provide a structure which will limit the local strain and stress to acceptable levels. Other examples of pressurised containers range from the common aerosol can, pneumatic reservoirs for lorry braking systems, oil storage tanks, spherical gas containers, submarine hulls, aircraft fuselages, water towers and silos. Pipework and pipelines appear in many industries and we might include with them penstocks for hydropower schemes although their complexity might suggest they be regarded more as pressure vessels than pipelines. However, they present some unique demands on welding technology in that in many situations the final welds joining sections have to be made from one side – the inside – of the pipe when it has been installed in a rock tunnel.

The design also has to recognise the possibility of fatigue cracking under any form of fluctuating load. In this context we cannot take refuge in the role of yielding or non-linear stress/strain behaviour to protect us from the effect of stress concentrations. Fatigue is dealt with at some length in Chapter 7 and is of particular significance in welded structures. Another design criterion used in some types of structure, such as building frames, is the deflection. If the floor joists bend too much the ceilings below may crack and if they feel too flexible the people in the building may feel unsettled. So, in addition to stress, the deflections caused in normal use may be a design criterion, usually set at some proportion of the span of the joist; a limit on deflection is typically 1/360 if there are brittle finishes such as plaster. Deflection or its corollary, stiffness, is also a design criterion in structures such as road vehicles, ships, aircraft, high masts, cranes and other types of mechanical handling equipment where not only excessive static deflection but resonance under fluctuating loads must be avoided. In machine tools and jigs stiffness may count for far more than stress in the design.

Limit state design

In the first part of the twentieth century improved knowledge about the statistical variations in load sources and material properties led to the development of a new design concept called *limit state* design. This approach is an alternative to the allowable stress method which, as we have seen, sets an arbitrary design stress (allowable stress) at a fraction of the yield or proof stress as a blanket device for covering all possible uncertainties. The limit state method of design envisages a number of states, the ruling one of which may be the ultimate limit state which is where the structure collapses. The design then ensures that the structure will not reach this state under the design loads. The various sources of uncertainty are identified and addressed according to their potential scale and effect. For example, in simple structural engineering the loads from different sources are factored depending on the reliability of their estimation and the likelihood of their occurrence. The stress to which the design is calculated can then be the yield stress but this does not mean that the maximum stresses developed in service will be significantly different from those used in the allowable stress method. In building construction the factors on different sources and combinations of loads are set to allow for the reliability of their calculation and the likelihood of their occurrence. These factors are called partial factors, for example on dead load the partial factor might be 1.4 whereas for an imposed load it might be 1.2. The design strength will be the minimum yield strength of the steel with a factor of 1.0, which for mild steel could be $275 \ N/mm^2$.

Serviceability limit states are those which affect owners or users of a building such as deflections, either static or dynamic; the latter are of interest in tall buildings which may sway under the effect of winds. Other serviceability limit states are those which require attention in maintenance programmes such as corrosion, wear and abrasion, accidental damage and fatigue cracks.

So to summarise:

• allowable stress method = actual loads with an arbitrary factor on strength.

- ultimate limit state method = (differently) factored loads and actual strength.
- serviceability limit state method = actual loads and actual behaviour.

Plastic design

The plastic method of design is a particular example of limit state design; it is of particular importance to the welding engineer because the concept requires potentially high integrity from the welded joints in terms of ductility and continuity. We saw in Chapter 3 how the stress in a steel beam in bending varied across its section so that the maximum stress was at the extreme fibres of the section. When this stress reaches yield our springboard will slowly bend down and become useless. However, it does not spell the end of the line for a beam with both ends fully built in or which is part of a frame in which the connections can transmit moments; indeed this yielding characteristic of steel is used to enhance the efficiency of building structures. So what happens in a steel beam when the yield stress is exceeded? We can return to the diagram of the stress distribution shown in Fig. 3.11.

As the bending moment increases from zero the maximum stress at the outer fibres of the beam approaches the yield stress, see line 1 on Fig. 4.1. If the moment is increased still more the outer fibres will begin to yield, see line 2, and the plasticity will spread further across the section, see line 3, also shown in Fig. 4.2. The beam is still capable of resisting the moment but eventually, if the moment is further increased, the whole of the section will have yielded and the section will then have developed its 'full plastic moment', see Fig. 4.3, and as a simple cantilever the loaded end will continue to fall and we may consider this beam to have failed as a structure.

The moment is then equal to the yield stress multiplied by the first moment of area of the section. In the case of a rectangular cross section we then have

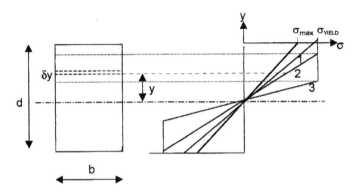

4.1 Stress distribution across a section under increasing bending moment.

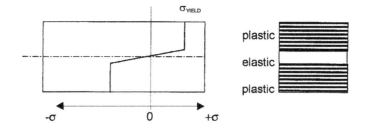

4.2 Development of the plastic region across a section in bending.

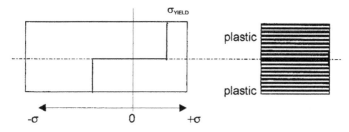

4.3 Wholly plastic section.

$$M = 2\sigma_{\text{YIELD}} \cdot b \cdot \frac{d}{2} \cdot \frac{d}{4} = \sigma_{\text{YIELD}} \, b \, \frac{d^2}{4} \qquad\qquad [4.1]$$

It will be seen that by analogy with the elastic modulus there is a plastic modulus

$$S_x = \frac{bd^2}{4} \qquad\qquad [4.2]$$

which is the same as the first moment of area of the section.

To show how this can come about we will use the same beam but instead of a single point load in the centre we will distribute the same load, *P*, uniformly across the span, see Fig. 4.4. The bending moment distribution is shown in Fig. 4.5. The derivation of these moments is quite complicated and outside the scope of this book and so the reader may either take them for

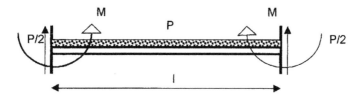

4.4 Beam with built-in ends under uniformly distributed load.

4.5 Bending moment distribution.

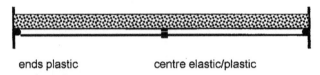

ends plastic   centre elastic/plastic

4.6 Two plastic hinges and the beam can still support the load.

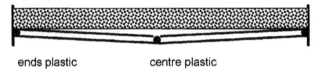

ends plastic   centre plastic

4.7 Three plastic hinges and beam can no longer support the load.

4.8 The beam becomes a mechanism and collapses.

granted or may prefer to consult one of the references on structures in the bibliography.

The point of including this case is to show that as the load P is increased the ends of the beam will yield first and become in effect like rather stiff hinges, called *plastic hinges* in structural engineering; once they have reached the full plastic moment the end moments will not increase as P increases, see Fig. 4.6. However, unlike the springboard the beam will not collapse at this stage of loading. It will continue to support the load as a beam with the ends being allowed to rotate. The moment at the centre of the beam will increase with increasing P and the beam will collapse only when the full plastic moment is reached at the centre, see Fig. 4.7, and the whole beam in effect becomes a mechanism, see Fig. 4.8.

The plastic method of design can be applied to all sizes of buildings and other structures in which there are elements where a number of plastic hinges can develop until a mechanism forms; use of this method can result in savings in the overall cost of construction. The significance of this approach to the welding engineer is that the steel beams and the joints at their ends and elsewhere need to have sufficient ductility as well as strength to develop the plastic hinges. This means addressing not only the ductility of the weld metal and heat affected zones but the overall ability of the joint to absorb the substantial deformations. Welding defects such as lack of fusion and penetration may become much more significant and the use of butt welds on backing bars may require close attention to root quality. The design and analysis of apparently simple single storey frames, such as are found in warehouses and supermarkets, is quite complicated and do not offer suitable examples here.

Summary

- The plastic method of structural design can be used only for steels with a clearly defined yield point and subsequent ductility.
- The structure collapses when there are sufficient plastic hinges to turn it into a mechanism.
- If the welded joints are to be become plastic hinges the mechanical properties and quality of the welded joints must allow full yielding to take place without fracture.

Joint types

Before thinking about the type of weld, we need to examine the type of joint which is to be made. It is helpful to place the commonly used welded joints into one of three types: the butt joint, the lap joint and the T joint. These are not rigorously defined terms but their convenience of application justifies the grouping.

The welded *butt joint*, see Fig. 5.1, is a joint in which two or more parts are joined end to end or edge to edge. The weld is within the outline of the final component and fuses the whole cross section.

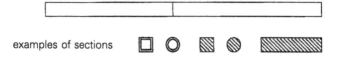

examples of sections

typical welding processes – arc, flash, friction, gas, electron beam, laser

5.1 Butt joint.

The *lap joint*, Fig. 5.2, is a joint in which one piece is placed over the other or between two others.

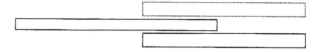

cross sections – sheet and plate and items built up from plate and sections

typical welding processes – arc, arc spot, resistance spot and seam

5.2 Lap joint.

In bolted or riveted structures a common type of joint is the butt joint made with butt straps, in reality a combination of lap joints, see Fig. 5.3.

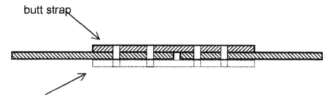

butt strap

second butt strap prevents secondary bending due to eccentricity of the joint

5.3 Butt strap.

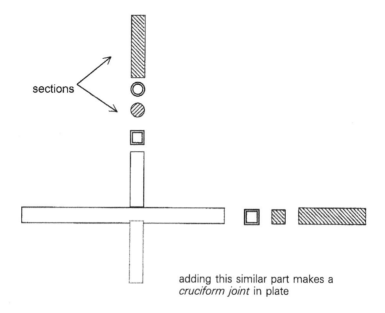

sections

adding this similar part makes a
cruciform joint in plate

Typical welding processes – arc, arc stud, friction.

5.4 T and cruciform joints.

This is used extensively in riveted structures such as airframes and can be seen in old ships and bridges.

The *T joint*, Fig. 5.4, is a joint in which the edge of one part is attached to the surface of the other.

Weld types

The types of joints shown in Fig. 5.1–5.4 can be made with a variety of welds, of which the two most commonly used are the butt weld, see Fig. 5.5, known in some quarters as a 'groove' weld and the fillet weld, see Fig. 5.6.

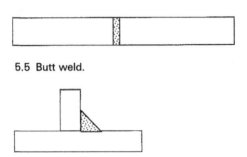

5.5 Butt weld.

5.6 Fillet weld.

Weld preparations

Thin plate and sheet can be arc butt welded with square edges as can thicker sections by using special welding processes and techniques. In conventional practice some shaping of the edges is required to be able to achieve controlled penetration whilst observing the constraints on welding conditions such as are imposed by the position, fit-up tolerances and the requirements for weld metal properties and weld quality.

In this context, *edge preparation* is the shaping of the parts to be welded and *weld preparation* is the combination of edge preparations fitted up to allow the desired type of weld to be made. The terms are most frequently applied to butt welds.

A conventional manual metal arc weld made in steel as a bead on plate will penetrate into the metal by only one or two millimetres and so any attempt to make a butt weld between the close square butted edges of steel sheet or plate of more than this thickness will leave lack of penetration, see Fig. 5.7. Trying to manual metal arc weld thin steel sheet of this thickness is not really practicable; more feasible everyday processes are gas or TIG and even then accurate cutting of the edges and firm jigging is necessary to produce a sound and consistent weld.

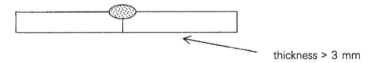

thickness > 3 mm

5.7 Bead on plate penetration.

Leaving a small gap between the edges will allow more penetration, see Fig. 5.8. This is very difficult to control and small variations in the gap left by cutting tolerances or by the thermal distortion opening and closing the gap can lead to the arc blowing through a larger gap and lack of penetration in a smaller gap. There is also the risk of lack of sidewall fusion where the

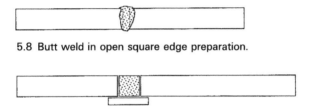

5.8 Butt weld in open square edge preparation.

5.9 Butt weld on backing strip.

arc does not consistently strike the sidewall. To avoid these problems the gap can be opened up and a backing strip can be placed behind the gap, see Fig. 5.9. A backing strip is usually of the same material as that being joined and is fused into the weld; it is usually left in place after the weld is completed. It can of course be removed by chipping and grinding but this is likely to leave damage to the weld root area which will then need repairing.

This open square weld preparation clearly uses a lot of weld metal; the more customary means of making a sound butt weld is by cutting a bevelled edge on the parts to be joined which leaves a single V shape to be filled, using less weld metal, see Fig. 5.10.

5.10 Single V preparation.

Again, a backing strip can be used or, in more sophisticated work, a backing *bar* which is usually made of a material such as copper, which has high conductivity, and can be water cooled or a ceramic. The backing bar can be shaped to give a neat weld root profile and is readily removed after welding is complete, see Fig. 5.11. Welding conditions must be chosen for the copper type to avoid fusing the backing bar into the weld metal. In automatic pipeline welding machines the backing bar can be attached to the line-up clamps. Another type of removable backing is a ceramic coated tape which supports the root and is removed after welding. When using the ceramic backing bar or tape for welding processes with fluxes such as manual metal arc or submerged arc, steps have to be taken to prevent the

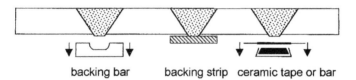

backing bar   backing strip   ceramic tape or bar

5.11 Various types of backing.

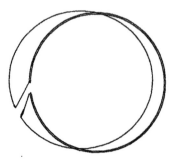

5.12 Sprung backing ring.

5.13 Attachment of backing strip showing alternative positions of tack welds.

molten flux from running under the weld pool so causing a poor root profile.

For butt welds in pipes both the backing strip and the ceramic backing can be made as a ring with obliquely cut ends which can be sprung into place and would then be self-supporting, see Fig. 5.12. The oblique ends allow the ring to adjust to the local pipe diameter whilst still remaining in contact with the pipe wall.

In medium to heavy fabrications the backing strip can be kept in place by tacking it to the back of the joint by intermittent or continuous fillet welds, see Fig. 5.13. Alternatively the tack weld can be made inside the preparation but care needs to be taken that the tack is either ground out prior to making that part of the butt weld or fully fused into the joint.

It will be seen that the butt weld on a backing strip allows a certain tolerance in fit up between the two parts and in this way is a useful device for absorbing shrinkage or growth as long as it is used under controlled conditions. In large or heavy tubes such as foundation piles the backing strip needs to be shaped so that a lead-in is available to prevent the backing strip being knocked off as the tubes are brought together. Further support will be offered if the strip is attached to any stab-in guide, see Fig. 5.14. However, the practicalities of lowering a pile weighing many tonnes sometimes means that the backing strip still gets damaged or knocked off altogether and the wise welding engineer will have a one sided welding procedure without backing in reserve.

In relatively thick walled tubes the backing can be made a part of the joint in what is called a spigot joint, see Fig. 5.15, but note that this is not a full penetration butt weld (see the section in this chapter on *partial*

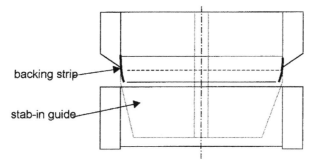

5.14 Backing strip in large tube to tube joint.

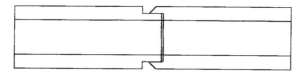

5.15 Spigot joint.

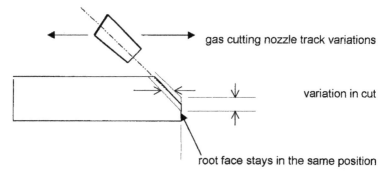

5.16 Tolerance of gas cutting on bevel.

penetration welds). This arrangement is self-locating and offers a reliable way of making a weld preparation.

A permanent backing may be undesirable for performance reasons such as fatigue, corrosion or interference with fluid flow in pipes and may also make the joint unsuited to straightforward non-destructive examination by radiography and ultrasonics. If no backing is to be used the bevel is not taken right through the thickness because the feather edge will melt away under the arc. A *root face* is left which will support the root of the weld, see Fig. 5.16. It also provides a tolerance in the event of the bevel cut not being straight; even if the cutting head wanders a little from the intended line it will still leave a constant *root gap* between the root faces as in the exaggerated example in Fig. 5.16.

Depending on the welding position the bevel may be cut on one or both

5.17 Weld preparation for joint in the horizontal-vertical position.

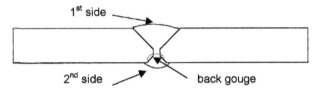

5.18 Back gouging in welding sequence.

of the edges to be welded. For joints in the HV position, such as are found in storage tanks and silos, it is convenient to have the bevel only on the top plate, albeit of a larger angle than would be used for a bevel on each edge. This gives a welding position closer to downhand which offers more flexibility in the choice of welding conditions, see Fig. 5.17.

A *single V* butt weld can be welded from one side only but this needs accurate fit-up and a highly skilled welder especially if welding has to be done in all positions. These requirements are satisfied of necessity in pipeline welding. Other users often seek a less onerous solution such as double sided welds, or at least a sealing run on the reverse side. In the latter case, if a weld free from root defects is required it may be necessary to back-gouge the root of the first side weld from the second side, see Fig. 5.18. Attempts to fuse the root completely with the first run from the second side without gouging or grinding can be unreliable.

In platework the single V preparation with its unequal disposition of weld metal about the central plane of the plate accentuates the thermal distortion which naturally occurs as the weld is built up run by run, see Fig. 5.19.

The distortion can be minimised by using a few large weld runs rather than many small ones. However this may not give the required weld metal

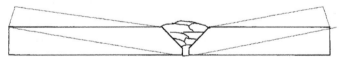

5.19 Distortion in one sided weld.

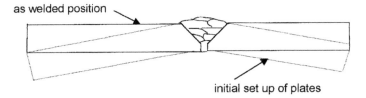

5.20 Presetting plates to reduce distortion.

properties and may not be feasible in any position away from the downhand position. Heavy clamping and dogging of the plates may minimise, but not eliminate, the distortion. In simple work it may be sufficient to preset the plates so that the distortion brings them into line, see Fig. 5.20.

Otherwise to minimise the distortion, particularly for thicker sections, a V preparation can be put on both sides of the joint giving a *double V* preparation, which to be effective is often asymmetrical in which case the smaller preparation is welded first, see Fig. 5.21.

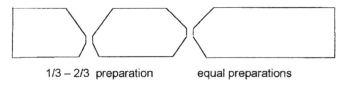

1/3 – 2/3 preparation   equal preparations

5.21 Double V preparations.

This two sided preparation has the disadvantage that the joint and the associated fabrication may have to be turned over to weld the second side. Another aspect of the choice of single or double sided weld preparation is the amount of weld metal. For the same thickness the single V uses more weld metal than the double V and weld metal is money, both in terms of the cost of consumables and energy but also in working time, see Fig. 5.22.

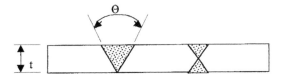

5.22 Volume of weld metal in weld preparations.

The area of the single V, see Fig. 5.23 (a), is

$$t.t. \tan \frac{\theta}{2} = t^2 \tan \frac{\theta}{2} \qquad [5.1]$$

and for the double V, see Fig 5.23 (b) the area is

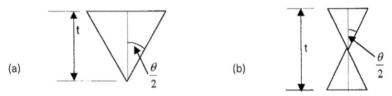

5.23 Weld preparation showing: (a) single V; (b) double V.

$$2 \cdot \frac{t}{2} \cdot \frac{t}{2} \cdot \tan \frac{\theta}{2} = \frac{t^2}{2} \tan \frac{\theta}{2} \qquad [5.2]$$

and so the volume of weld metal in the double V is half that in the single V.

The bevel is the simplest edge preparation because it can be flame cut, plasma cut or machined; in structural steels a double bevel with a root face is commonly cut in one pass using two or three gas cutting heads, see Fig. 5.24.

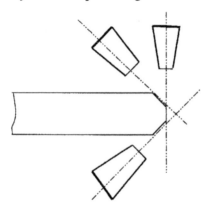

5.24 Single pass cutting of a plate edge with double bevel.

Another shape of edge preparation which is used to reduce weld metal and distortion is the J or U preparation. This gives a more even distribution of weld metal throughout the depth of the joint and may therefore give rise to less distortion across the joint. In the thicker sections in particular it can require less weld metal than the V preparation and the double sided version can be used in the same way as a double V.

The details of weld preparations are designed to suit the material, the joint configuration, the welding process (manual or mechanised), accessibility and the other features of the welding procedure. The variables in the preparations include those shown in Fig. 5.25.

Figure 5.26 shows a double U preparation. The features which may militate against the U preparation include the cost: it usually has to be machined rather than gas cut. For a plate edge the work has to be set up on a planer or mill which are rather slow devices for cutting a J; furthermore the size of plate which can be machined in this way is small compared with

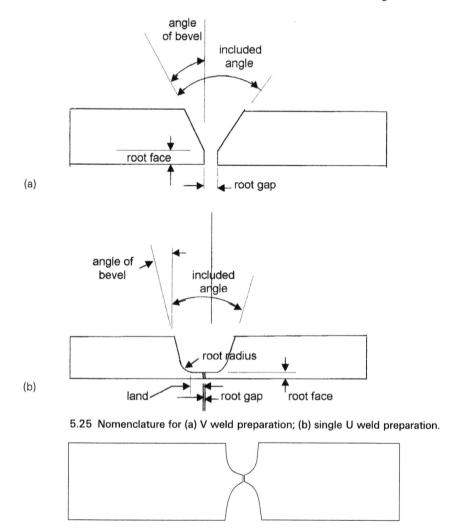

5.25 Nomenclature for (a) V weld preparation; (b) single U weld preparation.

5.26 Double U weld preparations.

the capacity of a gas cutting table. However, machining is not at such a
disadvantage for making edge preparations on tube and pipe as the work
can be turned on a lathe.

Another constraint on the use of the J preparation is the obstruction which
it can present to gas shielded welding nozzles in some joint configurations.
Depending on the relative sizes of the nozzle and the preparation opening, it
may not be possible to position the MAG gun without a large stick-out and/
or loss of shielding and possible lack of ability to fuse into the sidewall.
Figure 5.27 illustrates a J preparation for a T joint comparing access to the
weld root with a MIG/MAG gun and a stick electrode.

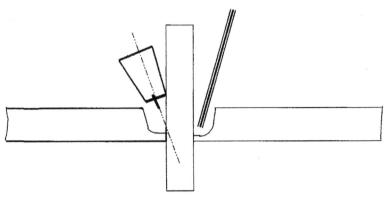

5.27 Limitation of access to MIG/MAG nozzle in J preparation.

The nozzle cannot get near the root, thus requiring a longer wire stick-out and consequently reducing arc voltage and disturbing the gas shield. There is insufficient space to manipulate the gun to achieve sidewall fusion. The result can be welds with lack of root and sidewall fusion and porosity. For certain types of work in steel it is possible to cut out a U preparation by placing the edges of two plates together and thermally gouging a groove using flame or air arc gouging.

The same principles for weld preparations apply when the two parts are of different thicknesses as in Fig. 5.28.

The actual dimensions of the preparation will depend on the material, the process and whether it is manual or mechanised, the position and the proximity of other parts as well as the consumables and welding conditions chosen to meet productivity and weld metal and heat affected zone

weld profile as required by design

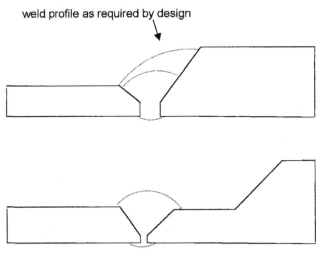

5.28 Alternative ways of joining parts of different thickness.

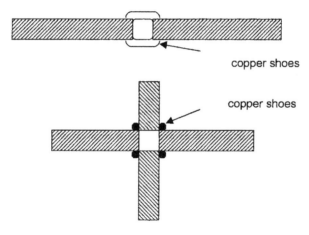

5.29 Preparations for electroslag welding.

properties. Various national and international standard specifications quote dimensions for weld preparations which by their nature are generalised and somewhat arbitrary. It is the welding engineer's knowledge of the factors referred to which should be the final basis of the choice of a weld preparation.

Specialised arc welding systems can be used to weld into narrow gaps. These have particular benefits for thick materials such as are found in process and power generation plant. Specifically designed weld preparations are used in place of the standardised weld preparations. Welding processes such as electroslag and consumable guide use a square edge preparation with a suitable gap for in-line butt joints and for cruciform joints, see Fig. 5.29.

The welding engineer will be aware that even the best finished edge preparation is of little worth if it becomes damaged or corroded in transit or storage. Paints may be used to prevent corrosion and it is common practice on linepipe and other pipework to fit protective plastic or metal rings over each prepared end.

Partial penetration butt welds

There are circumstances where the type of connection offered by the butt weld is required but where it is unnecessary for the weld to extend across the full section; in that case a partial penetration weld may be sufficient. Examples may be found in building columns or other members in compression and in lightly loaded joints. The partial penetration can be achieved in two ways: either by using a high current arc welding process onto a close square butted joint or by making a weld preparation with, in effect, a large root face, see Fig. 5.30.

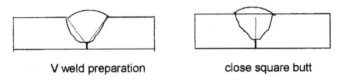

5.30 Weld preparations for partial penetration butt welds.

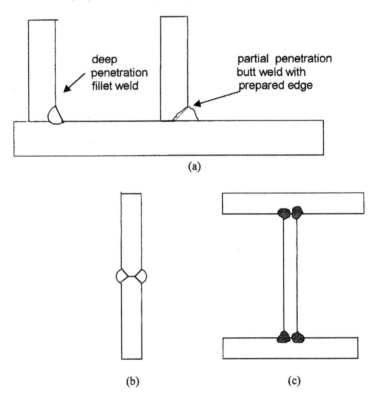

5.31 Examples of partial penetration butt welds.

Partial penetration butt welds can be used for T joints, see Fig. 5.31 (a), and it is possible to make double sided partial penetration butt welds, see Fig. 5.31 (b). Although these may be called butt welds and can be made into weld preparations they can be made as pairs of twin deep penetration fillet welds.

The beam and column sections made in this way, see Fig. 5.31 (c) are popular because they can be produced outside the standard range of hot rolled sections. The web is cut square and clamped to the flange, a partial penetration weld is then made with high current submerged arc welding operating under closely controlled welding conditions. It is a point for a rather academic discussion as to whether these welds should be defined as partial penetration butt welds or deep penetration fillet welds.

A cost benefit of using partial penetration welds arises not only from avoiding the need to back gouge the root but also in many cases because the joints can be self-jigging since no root gap has to be maintained during welding. However, a significant drawback of this type of weld is that the opportunity to assure weld quality by non-destructive examination by ultrasonic or radiography is effectively denied. A decision to use a partial penetration weld, as with a fillet weld, is therefore a recognition that quality assurance will be achieved entirely by process control, or an acceptance that the level of assurance required is lower than would be offered by a full penetration weld. The minimal scope for non-destructive testing of a partial penetration weld is also a source of cost saving. It is not unknown for fabricators to propose the use of partial penetration welds in place of the designed full penetration welds to reduce costs: the costs incurred not only in setting up, welding and inspection but also the costs incurred in weld repairs which might be consequent on the use of full non-destructive examination.

The radiograph, see Fig. 5.32, will not show the difference between a sound partial penetration weld and a partial penetration butt weld with a centre line crack in weld metal. The ultrasonic examination, see Fig. 5.33, probably will not either.

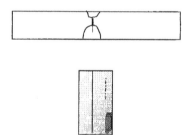

5.32 Radiography of partial penetration butt welds.

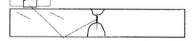

5.33 Ultrasonic examination of partial penetration butt weld.

It has to be recognised that the structural performance of a partial penetration weld will in some respects be inferior to that of a full penetration weld. This is particularly the case where fatigue life is a design criterion. Where low stress (brittle) fracture is of concern, steps should be taken to assess the capability of the weld metal, heat affected zone and parent metal of tolerating the presence of the unfused land. The assumed width of the land should allow for the possibility of lack of root fusion. Under any

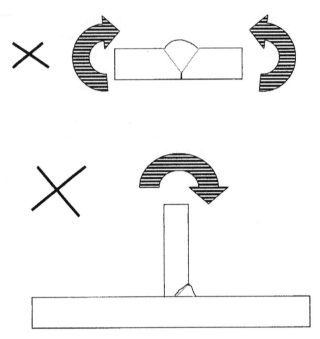

5.34 Inadmissible loading of joints.

loading condition the single sided partial penetration butt weld, as with the single sided fillet weld, should not be used in circumstances where the load puts the weld root in tension, see Fig. 5.34. The as-welded root is an area of crack-like features which may not tolerate strains which the sound parent material and weld metal will normally survive.

The edge weld

For a few special applications the edge weld is used, see Fig. 5.35.

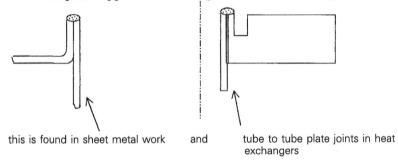

this is found in sheet metal work   and   tube to tube plate joints in heat exchangers

typical welding processes – gas, TIG, microplasma

5.35 Edge welds.

Making a butt weld in thin sheet requires very close fit-up and tightly controlled welding conditions which can be achieved only by the use of jigs with clamps, possibly with chill bars. For anything other than the simplest of straight line welds these will be quite complicated and expensive, added to which is the time taken to load and unload the jig. The edge weld, using a flange on one or both parts to be joined, offers a self-jigging arrangement which needs only a few simple locating guides or holes to hold the parts in the required relative positions whilst they are being welded. The welding process such as TIG or gas can be autogenous, i.e. with no filler, or with a filler wire.

The type of tube to tubeplate weld used in heat exchangers shown here is another form of edge weld. The tubeplate is usually very much thicker than the tube and if it is an alloy steel with strict weldability requirements it is difficult to get a heat input sufficiently low for the tube which is not too low for the plate material. The detail shown here is made by cutting a groove in the tubeplate which then leaves the same heat sink on both components for which the same welding conditions will be suitable. The detail is suitable for mechanised welding using, for example, TIG or microplasma in which an orbital welding head can be located off the bore of the tube and driven around the weld line. If the welding is done with the bore of the tube horizontal welding conditions can be automatically adjusted to suit the changing position. This detail is particularly useful if tubes have to be replaced; the weld can be cut off, the old tube removed and a new tube inserted and welded in the same way as the original. Non-destructive testing is limited to visual examination supported by magnetic particle or dye penetrant.

Spot, plug and slot welds

In resistance welding the preparation of the materials for the spot and the seam weld are basically the same. These joints are areas of two or more faying surfaces which have been locally fused either in a single spot or along a line. This book does not set out to cover resistance welding but examples are included because a structurally equivalent joint to this can be made with TIG or MAG process by melting through one sheet onto the surface of a second, see Fig. 5.36.

The nature of the resistance weld is such that the joints are usually lap joints but the flanged joint of the same type as shown for the edge weld is often used. Soft jigging to hold the parts together is of course helpful and the closing force of the electrodes finally brings and holds the parts together whilst the weld is being made. Nonetheless the parts have to be fitted up well or defective welds may occur. Some resistance welding equipment allows welds to be made effectively from one side only but none require special

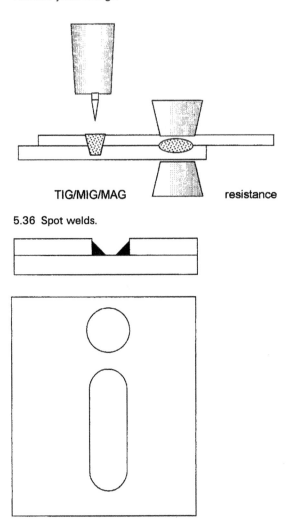

TIG/MIG/MAG   resistance

5.36 Spot welds.

5.37 Plug and slot welds.

preparation of the material beyond reasonable cleanliness and flatness at the joint. In contrast, TIG and MAG spot welds need clamps or hard jigging to keep the two parts firmly in contact whilst the weld is being made.

Plug and slot welds, see Fig. 5.37, are another device for connecting two plates in which a hole or slot is made in one plate and an arc weld made to connect the two.

For most purposes a fillet weld is made around the circumference of the hole or slot. The centre can be filled to make a full size weld which then equates structurally to the arc or resistance spot weld. This filling has to be done with care as a multipass weld. Just puddling in the weld metal is not acceptable since there is a risk of lack of fusion and leaving a crack at the final stop position in the weld. For this reason, various specifications accept only the fillet weld and place limits on the ratio of plug diameter to material thickness. As with the other spot welding techniques the two plates must be firmly clamped together before welding.

There are many other welding processes used for limited but important applications but it is not the purpose of this book to go into the design requirements for joints made with these.

Fabrication tolerances

In practice it will be unrealistic to assume that the weld preparations can be made exactly as required and, as with all engineering work, dimensional tolerances will need to be accepted. The need for tolerances arises from the following sources of dimensional variation:

- material dimensions, i.e. thickness, straightness, flatness
- material cutting tolerances
- fabrication tolerances of sub-assemblies
- thermal expansion during welding
- cumulative errors during erection

Some of these effects can be minimised by good cutting and fabrication practice but the welding procedure will have to accept some of the following features and tolerances must be set to ensure that the structure will have the performance required by the design.

- Mismatch, see Fig. 5.38, in thin plate can be corrected by clamping or dogging but doing this in thicker plate may lead to out of line fit-up as in Fig. 5.39.

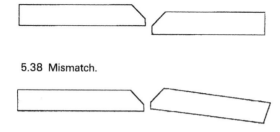

5.38 Mismatch.

5.39 Out of line (and out of round).

- Out of roundness (*roof topping*) in pipe, see Fig. 5.40, may be the result of poor rolling technique so that the ends of the plate are not rolled and have a short flat on them.

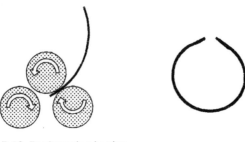

5.40 Roof topping in pipe.

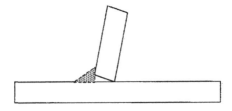

5.41 Ovality.

- The pipe may be oval, see Fig. 5.41. In line pipe (made specifically for pipelines) the ends of each piece are made round to within specification tolerances on diameter, wall thickness and ovality but the body of the pipe in between may be to a wider set of tolerances. This is acceptable for joining each piece end to end when laying a pipeline. However, when such pipe is used for structures whole lengths may not be suitable and cut sections within the length of the pipe may not match the ends or other mid-length sections; in this case fit ups need to be checked and corrected to ensure that unacceptable mismatch does not occur. Correction may be a simple matter of rotating one pipe relative to the other to obtain the closest match all round.
- If a there is a gap at the root of a fillet welded T joint either because the parts are not square or because the plate edge is not cut square, the fillet weld throat will be reduced as shown in Fig. 5.42, although this may be countered by a degree of root penetration. However, this cannot be relied upon.

5.42 Out of square.

- If the last joint in a large fabrication has the risk of an oversize root gap due to cumulative tolerances, see Fig. 5.43, it may be best to design a

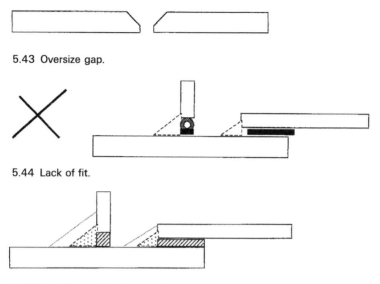

5.43 Oversize gap.

5.44 Lack of fit.

5.45 Possible repair schemes for lack of fit.

deliberately large gap into which can be fitted a closing length of beam or tube (a pup piece).

• Unauthorised 'fixes' for lack of fit, such as packing the gap with welding rods or bits of scrap and welding over them, see Fig. 5.44, should, of course, never be allowed! Any weld preparation which does not come within the specified tolerances should either be re-fitted or a repair scheme designed and approved by the relevant authority. For example, it may be allowable to pack a gap under a fillet weld provided that the weld size is increased to restore the throat size, see Fig. 5.45; this is always provided that the stresses are not magnified by increased eccentricity and provided also that the weld size does not become so large as to be impractical.

Attention to these matters is justified at the design stage as in the next example. The distortion created in sub-assemblies such as box girders, see Fig. 5.46, can cause mismatch problems in final assembly.

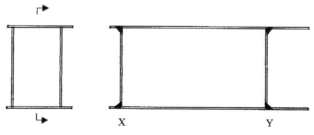

X   Y

5.46 Box girder.

5.47 Distortion of box girder flanges.

The upper and lower flanges distort unavoidably as a result of the welds attaching them to the transverse bulkheads. At end X the shortness of the flange overhang means that to line it up requires a severe bend, probably with plastic strain, and even then it may not be tangential to the horizontal. At Y the length of the overhang allows the end to be lined up by elastic bending of the flange plate until it is at the required line and level, see Fig. 5.47.

Residual stresses

The subject of residual stresses caused by welding is related to distortion since they both arise from the heating and cooling of metal during the welding sequence. The subject is briefly dealt with here since it has a bearing on the matters of fatigue cracking and brittle fracture. A very simple explanation of the reasons why stresses are set up by arc welding is shown in Fig. 5.48. This is of course a simplification and idealisation of the mechanism in which molten metal is sprayed or poured onto virtually cold parent metal from a source, the arcing electrode, which is moving, heating and melting a narrow band of the parent metal itself at the same time. The molten metal quickly freezes and in so doing fuses with the parent metal. The area in which this mechanism is operating at any moment is really quite small but, because of the thermal conductivity of the parent metal and the movement of the arc, effects are spread out over a large distance.

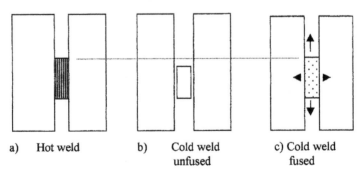

5.48 Origins of residual stresses in welded joints.

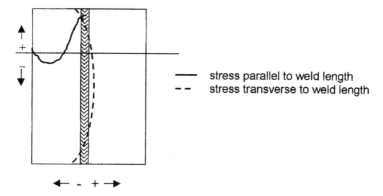

5.49 Typical distribution of residual stresses in a butt welded plate.

The hot solidified weld metal and the locally hot parent metal, see Fig. 5.48 (a), contract on cooling; if they were not fused to the cooler metal on either side they would contract as in Fig. 5.48 (b). Their fusion with the parent metal prevents such contraction, see Fig. 5.48 (c), and loads resisting this contraction are set up which are the source of stresses known as residual stresses. This is not the only source of residual stresses, for example the manufacturing process of steel plate sets up internal stresses because the surface of the plate cools before the core. In rolled sections the webs cool more quickly than the thicker flanges. When two plates are butt welded together a typical residual stress distribution develops as shown in Fig. 5.49. The residual stress in line with the weld can be up to the yield point in steels which accounts for certain aspects of fatigue behaviour described in Chapter 7. The level of residual stress is relatively lower in aluminium alloys because of the greater thermal conductivity which gives lower temperature gradients and stress peaks.

Access for welding

General considerations

Every welding process requires space around the joint for the equipment or heat source to reach the welding site. With the manual welding processes there is a need not only for access for the torch, rod or gun but the welder must also be able to see what the arc is doing whilst wearing a helmet or holding a shield. In a fully mechanised or automated welding facility there is no need for the human eye at the time of welding except perhaps to view what is happening during setting up. Nevertheless the equipment may be quite bulky requiring some space around the joint. With the beam welding processes, access may be required only for the width of the beam and this is exploited in some applications.

We saw earlier in this chapter how the design of weld preparations had to be suited to the welding process. The need for more general access can place some constraint on the design although the design concept should not be limited by this and the welding engineer may have to exert some ingenuity to solve the access problem presented by a novel design. Nevertheless there are some fairly basic features of the various processes which the design should recognise to prevent fabrication being more difficult, and therefore more expensive, than it need be. The following examples illustrate the potential difficulties.

Manual metal arc welding

In manual metal arc welding there are a number of mistakes which are commonly made in designing structures and structural details which prevent the welds being made properly. In the examples in Fig. 5.50 the rod cannot

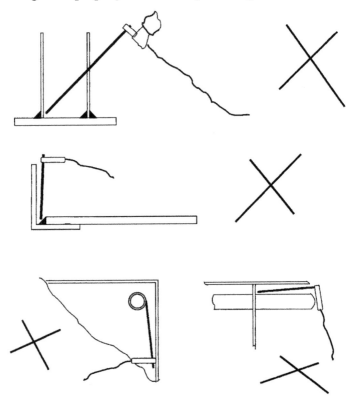

5.50 Typical details with lack of access to the weld root.

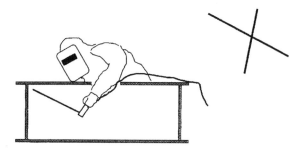

5.51 The welder may be able to reach the joint but cannot see the arc.

reach the weld root and in Fig. 5.51 the welder may be able to reach the weld root but cannot see it.

Manual MIG and MAG welding

In manual MIG and MAG welding the welder has to manipulate the torch, which has a much greater diameter than a welding rod, and support a wire feed conduit and a gas pipe and for high current work two cooling water tubes. There are the same access requirements as for manual metal arc but the difficulties are possibly accentuated. When welding within a closed space such as a box girder or tank the shielding gas can build up and replace the air so putting the welder at risk of asphyxiation unless a fresh air (**not** oxygen) supply is introduced. A build up of oxygen from cutting or shielding gas is equally dangerous as it can lead to flammable material such as clothing catching fire and burning fiercely.

Manual TIG welding

When using the manual TIG welding process with a filler the welder has to have both hands close to the weld, manipulate the filler wire and be able to see what is going on at the arc. There is no scope for arm's length welding here and access has to be very open.

Mechanised arc welding

MIG and TIG welding heads are quite compact and in making simple straight or circular welds a simple geometry is readily welded. Submerged arc equipment can be more bulky as the flux feed has to be accommodated.

A twin fillet weld can be made with mechanised gas shielded or submerged arc welding. Limits will have to be placed on the spacing of the stiffeners but the welding heads can be specially made to suit if necessary.

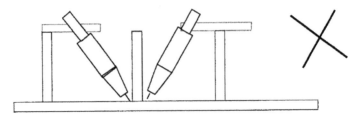

5.52 Clearance required for welding heads.

Note that flanges on the stiffeners may not be able to be accommodated, either before welding the stiffener to the main plate because they obstruct the welding heads, see Fig. 5.52, or after welding because there is no access to the underside of the flange for the reasons given above. This is an example of the method of fabrication requiring a compromise in the design; a flanged stiffener would generally be structurally more efficient. Welds can be made in a straight line or in a circular arc with simple moving heads or positioners. More irregular shapes can be welded using a mechanical follower system. Diaphragms or ribs at right angles to the stiffeners cannot be fitted before the longitudinal welding unless it is interrupted thereby detracting from the continuity of the welding with implications for quality control and structural performance. They can be welded in manually, or more complicated welding equipment can be devised to track the weld around the intersections if the design accepts the weld configuration. This is a type of construction for which robotic welding equipment is suited. Which route is followed depends largely on the equipment available or for which capital investment is allocated.

Welding information on drawings

An engineering drawing is an instruction as to the material, size and shape of parts and the way in which they are to be assembled and joined. Some drawings are termed project or conceptual designs and do not try to give very much detail. Others, developed from the foregoing, may be called design drawings and, as the product gets closer to manufacturing, there may be what are called shop drawings. The amount of information on any welded joint will differ between these sets of drawings. The conceptual design may show no joint details at all. The engineering design may well specify whether joints are to be made with fillet or butt welds because they may be selected for their structural performance. It is only at the shop drawing stage that information as to how the welds are to be made is likely to be put on the drawing. The way in which this is done and the extent of the information given differs depending on the practice in a particular industry or country. The information is most conveniently given in the form of symbols carrying reference letters or numbers; too much text on a drawing confuses the presentation. However, symbols are not always sufficient since they refer to other documents which may not be available to the person reading the drawing. This can be very inconvenient in the shop situation and partly for this reason many edge preparations, for example, are frequently to be found detailed on the drawings rather than indicated as a symbol.

The ISO symbols for welds on drawings given in ISO 2553 and EN 22553 represent a convenient system. The basic item in this system, as with its forerunners and other systems, is the arrow pointing to the weld upon which is hung information about the weld type, its size and the welding procedure to be used. There is a general purpose arrow, see Fig. 5.53, which in effect says that there is to be a welded joint here but its type is not yet decided.

5.53 Arrow indicating welded joint.

When information about the weld is available the standard type of arrow carries two horizontal lines: a full line called the reference line and a broken line called the identification line. The type of weld – fillet, butt, etc. – is denoted by a symbol on the arrow. A symbol sitting on the full line (see Fig. 5.54) refers to the arrow side of the weld and a symbol on the broken line (see Fig. 5.55) refers to the weld on the other side. Figure 5.56 shows these sides.

5.54 Fillet weld on 'arrow' side.

5.55 Fillet weld on 'other' side.

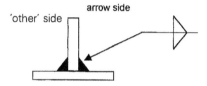

5.56 'Sides' of a joint.

If the two welds are of the same type and size, i.e. a symmetrical joint, the identification line can be omitted as in Fig. 5.56. The weld size is placed to the left of the symbol and for fillet welds it is necessary to indicate whether it is the throat thickness (a) or the leg length (z). The length of the weld is put at the right hand side of the arrow. Figure 5.57 indicates an 8 mm throat fillet weld 300 mm long on the arrow side of the joint.

5.57 Dimensions of a weld.

The type of welding process to be used may be inserted in the symbol (see Fig. 5.58) using the process reference numbers from ISO 4063.

5.58 Welding process identifier (manual metal arc).

The number of the welding procedure specification can be inserted adjacent to the symbol as in Fig. 5.59.

5.59 Welding procedure specification on the arrow.

In this case we presume that since the welding procedure specification will include the welding process it will be unnecessary to identify the process. Welding all round a joint is shown in Fig. 5.60 by a circle and by a site weld in Fig. 5.61.

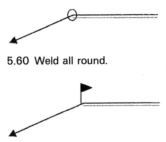

5.60 Weld all round.

5.61 Site weld.

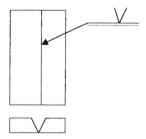

5.62 Symbol for a single sided V-preparation butt weld.

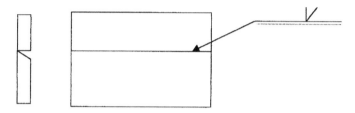

5.63 Arrow pointing to the prepared edge.

The butt weld is indicated by one of a number of symbols as in Fig. 5.62. If the edge preparation is to be on only one part then the arrow points to the prepared edge, see Fig. 5.63.

ISO 2553 lists a large number of pieces of information which can be attached to the basic arrow covering a wide range of joint and weld types and welding processes. Information about inspection and other activities is also allowed for. For a description of some of the various weld symbol systems in use readers are advised to consult *Weld symbols on drawings* (see bibliography).

Summary

- Most arc welding processes require edge preparation.
- More weld metal = more cost and more distortion.
- Partial penetration welds may be useful sometimes but do not rely on the throat thickness without a high level of control.
- Allow fabrication and assembly tolerances.
- If a welder cannot see or reach a joint he cannot weld it.
- Standard weld symbols are a concise way of conveying weld requirements.

6

Calculating weld size

Butt welds

Full penetration

The strength of a full penetration butt weld is the strength of the cross section of the material in the joint. The material defining the strength can be weld metal, the heat affected zone or the parent metal. In the commonly used structural steels the weld metal is usually stronger than the parent metal. The heat affected zone could well be stronger but its width is usually so small that it does not influence the strength of the whole joint. In higher strength steels it is possible to obtain weld metals of lower strength than the parent metal and the heat affected zone may be weaker than the parent metal.

In the common carbon and carbon manganese steels with yield strengths up to, say, 400 N/mm^2 most arc welding consumables will give a yield and tensile strength matching or exceeding the parent metal. Strength calculations for static loading then do not have to recognise the presence of full penetration butt welds at all. In the case of lean steels deriving their strength from thermo-mechanical treatment the results of welding procedure tests will need to be examined to see if the weld metal or, more likely, the heat affected zone display strengths below the strength of the parent metal. If so, some allowance may have to be made in what stresses are acceptable for the design cases. In very high strength steels, e.g. with a yield strength of over 500 N/mm^2, if undermatching weld metals are used then a reduction in any design stress based on the parent metal strength may be necessary. Full penetration butt welds in most austenitic and ferritic stainless steels can be treated as having the static strength of the parent material. The conductivity of aluminium and its alloys and their metallurgical behaviour cause them to have a much wider weld heat affected zone than steel so that its properties

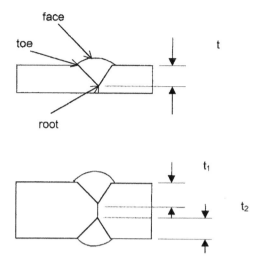

6.1 Dimensions of partial penetration butt welds.

can influence the behaviour of the whole joint. In the weldable aluminium alloys both the heat treated and the work hardened varieties suffer from local deterioration in their strength adjacent to the weld metal which is recognised in the static design stresses for welds which are used for these alloys.

Partial penetration

The partial penetration weld, see Fig. 6.1, is a convenient device (see Chapter 5) but it is of indeterminate size unless sophisticated non-destructive examination techniques are used – in which case we might ask why not use a full penetration weld in the first place? We call it a butt weld but in behaviour and configuration it has much in common with the fillet weld. Like the fillet weld it has a face, a root with an unknown and possibly erratic profile and a throat which is indeterminate because we cannot see the shape of this erratic profile.

The weld throat size, t or the sum $t_1 + t_2$, is used as the load carrying cross section under axial or shear load. Various national standards make allowance for the possibility that there may be lack of fusion at the root by subtracting a fixed dimension or a proportion of the weld throat from the idealised throat. For axial loads the allowable tensile stress in the parent material is used across the net section so derived and the corresponding allowable shear stress for shear loads along the length of the weld. In compression, provided that the root faces (or unfused lands as they may be called where the welds are small in relation to the thickness) are flat and

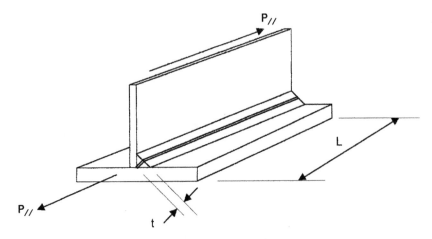

6.4 Fillet weld with longitudinal shear load.

The weld throat shear stress is

$$\tau_{//} = \frac{P_{//}}{Lt} \qquad [6.1]$$

This stress is one of two types and two directions of stress which are postulated to exist in a fillet weld. Two are suffixed with the sign for parallel, //, indicating that they result from a load parallel to the length of the weld. Two are suffixed with the sign for perpendicular, ⊥, indicating that they result from a load perpendicular to the length of the weld, see Fig. 6.5. These are merely symbols describing the type (normal and shear) and direction of stress and do not represent a set of internally balanced stresses.

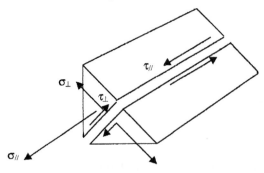

6.5 Stress notation for fillet weld.

Tests on fillet welds in mild and high yield steels with nominally matching weld metal found the normal stress, $\sigma_{//}$, to have no measurable effect on the strength of the weld. This type of stress is most common in the web to flange weld in an I beam in bending. For design purposes in structural steels it was

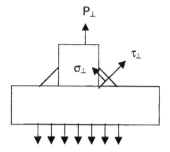

6.6 Twin fillet welded joint under load.

found that the three other stresses could be related to an allowable stress by a formula of the type

$$\beta\sqrt{\sigma_\perp + 3(\tau_\perp^2 + \tau_{//}^2)} \leqslant \sigma_c \qquad \text{[6.2]}$$

and

$$\sigma_\perp \leqslant \sigma_c$$

where σ_c can be the allowable tensile stress or limit state stress. This is used as a basis for fillet weld strength in a number of standards in which values for ß are typically in the region of 0.8–0.9 depending on the strength of the parent metal.

The example above shows how $\tau_{//}$ is calculated, another simple example will show how the other two stresses are derived.

Figure 6.6 shows twin fillets, each with throat thickness t. Then, resolving vertically

$$P_\perp = \frac{2tL}{\sqrt{2}}(\sigma_\perp + \tau_\perp) \qquad \text{[6.3]}$$

and horizontally we can see that

$$\sigma_\perp = \tau_\perp \qquad \text{[6.4]}$$

and so

$$\sigma_\perp = \tau_\perp = \frac{P}{2\sqrt{2tL}} \qquad \text{[6.5]}$$

The stresses so calculated can be put into the equation with the relevant parameters ß and σ_c to arrive at a value of t for the design. If there is a load creating a parallel shear stress then this stress can also be entered into the equation.

This is a rather cumbersome procedure for run-of-the-mill work and it is often customary to use just the load divided by the weld throat as a measure of weld stress which in structural steels is then compared with the allowable or limit parent metal shear stress. If a parallel load is present then the two throat stresses are summed as a resultant square root of the sum of the squares.

If τ_t is the nominal fillet weld throat stress then

$$\tau_t = \frac{\sqrt{P_{//}^2 + P_\perp^2}}{2Lt} \qquad [6.6]$$

For other materials more complicated routes may be used and the relevant standard or code of practice is followed.

Calculating the strength of welded joints

A few examples will show how the strength of fillet welded joints is calculated.

a) A flange welded to a shaft, see Fig. 6.7

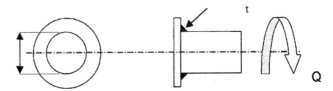

6.7 Flange welded to shaft.

Weld throat t, torque Q, weld length $= \pi d$, throat area πdt, a lever arm of $d/2$
and so

$$Q = \pi dt \, \tau_t \, \frac{d}{2} \qquad [6.7]$$

or

$$\tau_t = \frac{2Q}{\pi d^2 t} \qquad [6.8]$$

If the weld throat is 6 mm and the shaft diameter is 100 mm then with an allowable weld throat stress of 115 N/mm^2

$$Q = 115 \times \pi \times 10\,000 \times 6/2 \text{ N mm} = 10.8 \text{ kNm}$$

b) A rectangular bar or box section fillet welded to a column, see Fig. 6.8.

The weld throat size is t. We need to find the stress in each part of the joint which we do by treating the weld throat profile as we would a hollow section, see Fig. 6.9.

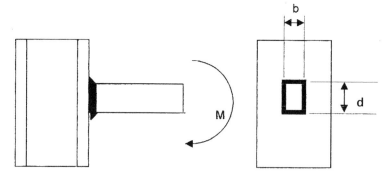

6.8 Bending load on a rectangular bar or box section welded to a column.

6.9 Weld throat profile.

The equivalent to the wall thickness is the weld throat thickness, t, and we proceed to calculate the section bending modulus. We are unlikely to give rise to gross inaccuracy if we take the centre line of the weld throat as having the same dimensions as the attached bar, i.e. $b \times d$.

Taking dimensions $b = 120$, $d = 300$, $t = 5$ and $M = 50$ kNM

Item	Area	Dims mm	A	y	y^2	$Ay/10^3$	$Ay^2/10^6$	$I/10^6$
Top	$b.t$	120 × 5	600	150	22500	90	13.5	0
Sides	$d.t$	300 × 5(2)	3000	0	0	0	0	22.5
Bottom	$b.t$	150 × 5	600	−150	22500	−90	13.5	0
Total	–	–	4200	–	–	0	27	22.5

I for the whole section $= 49.5 \times 10^6 \text{mm}^4$ and so $Z = I/150 = 0.33 \times 10^6 \text{ mm}^3$

Then the maximum throat stress $= M/Z = 50 \times 10^6/0.33 \times 10^6 \approx 152 \text{ N/mm}^2$

c) The same principle can be applied to other profiles. We then need to look at the case where there is shear load and a bending moment, see Fig. 6.10. Take the same moment and add a shear load S.

The shear can be taken by all of the welds and so the shear stress due to the shear load of 100 kN is

$100\,000/4200 = 23.8 \text{ N/mm}^2$

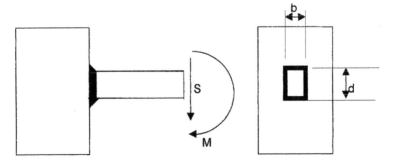

6.10 Shear and bending loads.

In the top and bottom and the top and bottom of the side welds this can be added by the simple method, i.e.

$$\text{stress} = \sqrt{23.8^2 + 152^2} \approx 154 \text{ N/mm}^2$$

If the attached item were not a bar but a thin walled hollow section we could assume all the shear to be taken in the side welds. The shear stress in the side welds would then be

$$100\,000/3000 = 33.3 \text{ N/mm}^2$$

and then the maximum stress in the side welds would be

$$\sqrt{33.3^2 + 152^2} \approx 156 \text{ N/mm}^2$$

Methods of measuring stress

There are no methods of measuring stress in the practical site situation. What can be measured is strain and there are a variety of techniques which can be used. The choice of the techniques depends on the scale of the structure, the level of detail of the stress measurement, the environment for both installation and service and the time over which measurements are to be made.

Probably the most common type of strain measuring device is the electrical resistance strain gauge. This consists of a number of fine conductors which are bonded to the structure; they stretch or compress in sympathy with the surface to which they are bonded. Their change in length causes changes in their electrical resistance which is detected by having the gauges connected into a balanced electrical bridge circuit. Changes in the resistance of the gauge unbalances the bridge; the change in resistance of one of the complementary bridge arms required to re-balance the bridge is a measure of the strain change. The installation of these gauges requires a finely prepared surface and careful bonding followed by waterproofing if they are in a location exposed to any level of moisture. The circuits must incorporate compensation for temperature changes which can otherwise produce spurious readings. Strain gauges can be bonded to the structure as a permanent monitoring installation or used as a short term laboratory tool. They are made in a wide range of sizes either as individual gauges or in multiple arrays for measuring strain gradients and in a variety of configurations, for example in a circular format for measuring strains at the periphery of holes. Electrical resistance strain gauges are used as permanent installations in some types of load cell and weighing devices.

Another form of strain measuring device detects the change in the natural frequency of a wire when its tension changes. This gauge was originally developed for measuring strains in building frames and is quite large but robust, an important feature for site installations. Optical interference phenomena are used for measuring surface strains in high resolution laboratory work. A microscopic grid is etched or engraved on the surface and changes in the dimensions cause the movement of interference fringes when viewed through an optical system.

Another method requiring close control of the environment is to spray a brittle lacquer onto the surface. This dries to a very thin film which cracks when a certain level of strain develops.

Summary

- For calculation purposes welds are butt welds, partial penetration butt welds and fillet welds.
- Butt welds reproduce the whole cross section of the member and in structural steels no calculation of their static strength is required.
- Partial penetration welds have a design tensile strength proportional to their total throat. In some applications the full member section is used in calculating the design compressive strength.
- Fillet weld strength is calculated on the basis of the weld throat.
- Deep penetration fillet welds may be assumed to have a larger throat provided that there is evidence of this throat being achieved.

Problems

6.1

A 75 mm diameter tube through a 25 mm thick plate acting as a lever on a shaft. The tube is fillet welded to the plate on both sides.

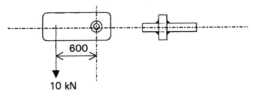

What weld size is required if the throat stress is not to exceed 120 N/mm^2?

Fatigue cracking

Load histories

Many parts of machines, plant and vehicles operate under fluctuating loads. The fluctuations may be slow, as in the tidal sequence of the load on a dock gate, or they may be fast as in a vibrating screen for separating and grading gravel, coal or other solids. The stresses may be induced by externally applied loads or they may be due to resonance at the natural frequency of a structure.

It is convenient to adopt a standard nomenclature for the quantities we have to consider, see Fig. 7.1. Since we are dealing with load or stress as a function of time, we can make a graph to show the variation of load with time, or *load history* as it is called. Later on we shall find that we appeal to stress rather more than load, so from now on the stress will be the variable of interest and the stress history is a prime piece of information.

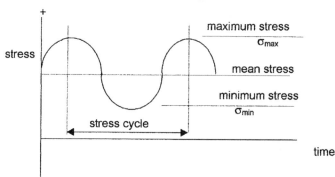

7.1 Nomenclature for a stress history.

Stress range is the difference between the maximum and minimum stress in the cycle.

$$\text{stress range} = \sigma_{max} - \sigma_{min} \qquad [7.1]$$

Another term is *stress amplitude*, which is half the stress range and owes its origins to the description of harmonic motions. It is not used in the calculations but occurs in qualitative terms such as *variable amplitude* which appears later in this chapter. *Stress ratio*,

$$R = \sigma_{min}/\sigma_{max} \qquad [7.2]$$

So if the minimum stress in a cycle is zero then $R = 0$, a pulsating cycle; if the stress changes from a tensile value to the same value in compression then $R = -1$, an alternating cycle.

Weld features

Arc welds are a complex mixture of metallurgical structures set in an erratic physical shape. The weld toe is the boundary between the once molten and rapidly frozen mixture of molten parent metal and added consumable and the still solid parent metal. The latter has undergone its own severe heating and cooling cycle and will not have retained its original metallurgical structure. The mechanisms involved leave at the toe of the welds in steel very small crack-like features, named intrusions by their discoverers, typically of around 0.15 mm depth which are not detected by conventional non-destructive examination techniques, see Fig. 7.2. These are not what is commonly called undercut – a much deeper but more rounded feature at the toe of the weld.

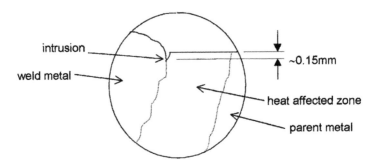

7.2 Diagram of weld toe detail.

Fatigue crack growth

Fatigue cracking is the step by step growth of cracks driven by fluctuating stresses which may be much smaller than the tensile strength and the yield or proof stress. A fatigue crack starts at a point of high local stress concentration such as an intrusion and the rate of growth of such cracks is a function of the *stress intensity* at their tip. The *stress intensity factor, K,*

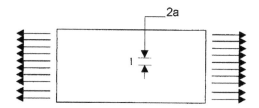

7.3 Crack in plate transverse to load direction.

is a parameter used in fracture mechanics which defines the redistribution of stress in a body arising from the introduction of a crack. The definition of K for any particular geometry of crack and loading condition is a matter for stress analysis and there are standard solutions for many types of configuration.

A simple example, see Fig. 7.3, is a through thickness crack of length $2a$ in an infinite plate under a uniform stress σ acting at right angles to the plane of the crack.

For this situation

$$K = \sigma\sqrt{\pi a} \qquad\qquad [7.3]$$

This has the units of stress $\times \sqrt{}$ length and it can be seen that the same value of K can result from a high stress and a short crack or from a low stress and a long crack. For other configurations a correction factor can be applied to this expression which becomes

$$K = Y\sigma\sqrt{\pi a} \qquad\qquad [7.4]$$

The values for Y are published for various configurations including welded joints.

Under a fluctuating stress

$$\Delta K = K_{max} - K_{min} \qquad\qquad [7.5]$$

The crack tip is stretched by each cycle of stress and provided the value of ΔK is above the threshold (see below) it will extend into the material by a small distance each time it is stretched, see Fig. 7.4.

The amount by which the crack extends with each cycle of stress is given by a relationship between ΔK and the rate of fatigue crack propagation in the form

$$\frac{da}{dN} = C\,(\Delta K)^m \qquad\qquad [7.6]$$

in which a is the crack length, N is the number of load (or stress) cycles, C and m are constants applicable for certain materials, environments and

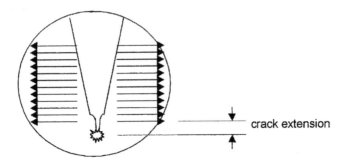

crack extension

7.4 Crack tip extension detail.

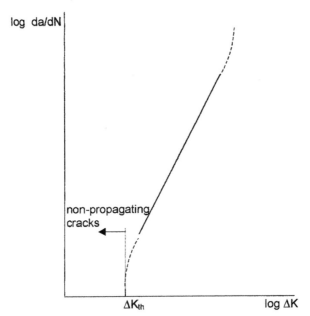

7.5 Graph of crack growth rate da/dN against ΔK.

stressing conditions. When plotted on a log log scale this would be a straight line as in Fig. 7.5.

However, in practice, when laboratory tests are made it is found that there is a certain value of ΔK below which the crack will not start, sometimes referred to as a threshold, ΔK_{th}. At the other extreme, high values of ΔK approach the critical value at which brittle fracture or gross plastic strain occurs. These features are indicated on the graph in Fig. 7.5. The intrusions in the weld will act as the initial crack so that provided that the value of ΔK is above the threshold value the crack will start to grow with each stress cycle. We have assumed up to this point that the stress cycles are uniform, i.e. of constant amplitude. As the crack grows, the value of ΔK

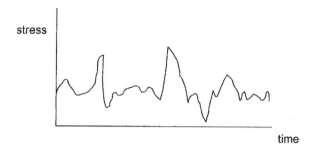

stress

time

7.6 Variable stress history.

increases and so the rate of crack growth increases. What we see in practice is that the crack remains very small for a large part of the life of the item but finally grows very quickly to a size at which it can be seen with the naked eye. Many industrial products are painted and this can obscure the growing crack until its continued opening breaks the paint film. Even then the crack may not be obvious to the untrained eye until perhaps rust or grease accumulates in the crack. For these reasons, once a fatigue crack is seen by eye a large part of the life of the item will probably have been used up.

Many products operate not under a constant stress amplitude but under a variable stress amplitude, see Fig. 7.6. To begin with, the smaller stresses in the history may give ΔK below the threshold and will not extend the crack and only the larger stress will cause the crack to grow. As the crack gets larger the smaller stresses will start to give ΔK above the threshold and so more of the stress cycles will cause crack growth and the crack will grow at an ever increasing rate.

The constants C and m were described as being material and environment related constants. In some environments not only is the rate of crack growth enhanced by their effect but the threshold value of ΔK may reduce or vanish altogether. This effect is probably most commonly encountered in steel structures in seawater where steps are taken to counteract the effect by imposing electrical currents which also prevent corrosion of the surface. The electrical currents are produced by attaching light alloy bars to subsea structures or by passing electrical current through anodes attached to the structure. Such protection is not effective for parts of the structure which are within the tidal area where they experience alternately wet and dry conditions. For those parts of the structure a paint system has to be used.

The practical design approach

The concepts of fracture mechanics help to explain the way in which fatigue cracks grow but they require refined stress analyses to enable them to be applied to a real structure using appropriate parameters with levels of confidence established in the significance of each result. This is therefore not the technique used for normal everyday design activities for which use is made of graphs of welded joint fatigue life against stress. These are known as SN curves, S being a symbol commonly used for stress and N, as seen above, denoting the number of cycles. The method is based on data obtained in many hundreds, or thousands, of laboratory fatigue tests on a range of different types of welded joints. We shall see later that welded joints are a special example of material as far as fatigue cracking is concerned.

Fluctuating stresses applied to a metal specimen under test conditions show that there is a relationship between the magnitude of the stress fluctuation and the number of cycles which the specimen survives before it fractures. The stress can be applied by a bending load or by an axial load, see Fig. 7.7. Early researchers favoured the bending load because it can be easily applied with a simple rig driven by an electric motor. This could run up in a relatively short time the large number of cycles necessary for some tests. Axial loading is more difficult to contrive but as we shall see it can apply a range of loading conditions which the rotating machine cannot.

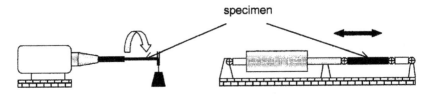

7.7 Rotating bending and axial load fatigue testing machines.

The rotating test machine places an alternating stress on any point on the surface of the specimen. When that point is at top dead centre it experiences a tensile stress and as it rotates the stress decreases to zero after a quarter of a turn and then increases to maximum compressive stress at bottom dead centre, see Fig. 7.8.

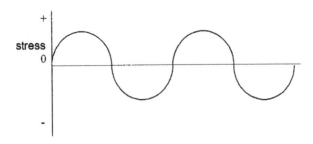

7.8 Alternating constant amplitude stress history.

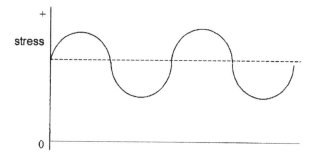

7.9 Constant stress amplitude history with tensile applied mean stress.

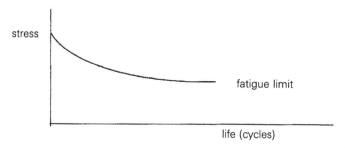

7.10 Typical SN curve for plain specimens.

We saw earlier in this chapter that for this load history the stress ratio, R, $= -1$. With the axial load machine we can impose a *mean stress* about which the stress can be made to fluctuate. In Fig. 7.9, if $\sigma_{max} = 2\sigma_{min}$, $R = 0.5$.

This is an important feature because it is found that if several specimens are tested at the same stress range but at different mean stresses the fatigue life gets shorter as the mean stress gets higher. The final fracture occurs when the crack becomes so extensive that the remaining cross section is insufficient to support the load. The number of cycles at which this occurs is called the *fatigue life* of the joint. When a number of specimens have been tested, each at a different stress, the results can be plotted on a graph, the SN curve, which will look something like that in Fig. 7.10.

A large part of the life is spent in what is called *initiation* of the crack. This is an event which starts on a molecular scale as various movements take place inside the material which eventually lead to the growth of a micro- and then macroscopic crack. Below a certain stress range, called the *fatigue limit*, the specimen will not crack at all. This stress range is quite high, in steels it is typically 50% of the yield stress. Any surface feature which gives rise to a stress concentration will cause the life to be shorter; in effect the material local to the concentration is working further up the SN curve. At this point we move away from plain steel or other metals and move on to

7.11 Typical crack location in transverse butt weld.

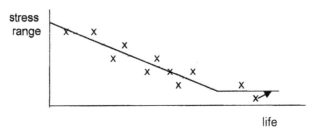

7.12 Typical SN curve for welded joint.

consider welded joints for which there is a different pattern of results. The simplest welded joint by which to explain the relevant test data is a butt joint in plate made with full penetration butt weld which is subjected to a fluctuating axial load.

The effect of the residual stresses in the welded joint is to put the mean stress near to yield and any difference in the applied mean stress has no significant effect. So the figure we now plot on the stress axis is the stress range and if the results are plotted on a log log scale the curve reduces to a straight line. So instead of having to use a whole nest of SN curves, each for a different R, we only have to use one. There is a further unique feature of welded joints already referred to above and that is the presence of weld toe intrusions. When the weld axis is transverse to the direction of the applied stress these intrusions will induce a fatigue crack to propagate from a weld toe, see Fig. 7.11. Their further effect is that they can represent an immediate crack starter so that there is no initiation phase and a very much lower 'fatigue limit' than for plain material, see Fig. 7.12.

In practice each weld and each specimen will be slightly different and the results (x on Fig. 7.12) show some *scatter* and do not exactly lie on the line. The arrow on the lowest result means that the specimen had not broken at that life and the test was stopped. If a specimen comprising a plate with fillet welded tabs is tested, see Fig. 7.13, the result might be considered surprising

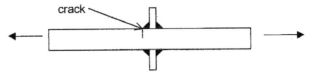

7.13 Fatigue crack location at fillet weld toe.

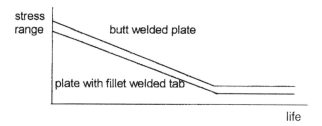

7.14 SN curve for two weld details.

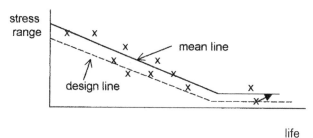

7.15 Derivation of design SN curves.

for the fatigue life of this specimen, in which no load passes through the weld, is lower than that of the loaded butt weld, see Fig. 7.14.

The reason for this apparently anomalous behaviour is that the shape of the fillet welded tab causes a stress concentration to occur in the plate at the weld toe (see Chapter 1) which adds to the stress experienced by the metal at the weld toe intrusion. This causes a larger ΔK than at the toe of the butt weld for the same load giving a higher value of da/dN, at least to start with. Tests have shown how various types of weld joint configurations give different SN curves. These are published in design data and standards as a range of joint types categorised by their fatigue behaviour.

These test results display the scatter referred to above and the actual design data have been derived for each category not by taking a mean line through the results, which would mean that half the joints would fail prematurely if they were designed for that life, but by taking a lower line which includes a large proportion of the results. In statistical terms this lower limit, or design SN curve, is drawn so that it gives a 97.5% probability of survival. The mean line gives 50% probability of survival, see Fig. 7.15.

Complex load histories

So far we have been using the results of constant amplitude fatigue tests to design components themselves operating under constant amplitude loading. This is not the case for many products which experience variable amplitude

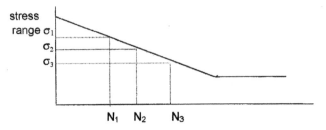

7.16 SN data for cumulative damage calculation.

load histories. All types of vehicles – whether land, sea or air – mechanical handling equipment and other plant experience variable amplitude loading. A method of using the standard design data based on constant amplitude test results to design for variable amplitude loading is known variously as the *cumulative damage rule, Miner's Rule* or the *Palmgren-Miner Rule*. In principle this is very simple although in practice it can be quite difficult to derive the necessary input. The basic step is that the constant amplitude SN curve appropriate to the joint detail is selected.

The various stress ranges in the load history are counted; to keep it simple we will take as an example a history with only three stress ranges, see Fig. 7.16, with the number of cycles of each stress range being n. The rule says that the fraction of the whole fatigue life used up by each stress range is the number of cycles at that stress range, *n*, divided by the number of cycles at that range which would be required to reach the design line, *N*. So the amount of life used up by any one stress range is n/N. This is called the *damage*. We then add the damages from all the stress ranges to form the cumulative damage (hence the 'cumulative damage rule'). When this cumulative damage reaches 1 the life of the item is deemed to have expired. In mathematical terms we say that for survival

$$\Sigma \frac{n}{N} < 1 \qquad\qquad [7.7]$$

In our example

$$\frac{n_1}{N_1} + \frac{n_2}{N_2} + \frac{n_3}{N_3} < 1 \qquad\qquad [7.8]$$

As an example take a Class F detail on an overhead travelling crane at a reinforced concrete plant. The crane does three basic lifts as follows:

Movement	Load, tonnes	Lifts/year	Stress at weld detail N/mm^2	Constant amplitude life, cycles
1. Lift and carry full skip to casting bay	10	6000	80	1 400 000
2. Lift and transport empty skip back to batching plant	1	6000	8	> 10^8
3. Lift finished RC beam on to transporter	25	1800	200	80 000

So for one year's operation the damage is

$$\Sigma \frac{n}{N} = \frac{6000}{1\,400\,000} + \frac{6000}{10^8} + \frac{1800}{80\,000}$$

$$= 0.00429 + 0.00006 + 0.02250$$

$$= 0.02685$$

If that is the damage for one year it will take

$$\frac{1}{0.02685} = 37 \text{ years}$$

for the damage to reach 1.0 at which the fatigue life is used up. Experiments have shown that in many circumstances this rule is very conservative and in welded details values of up to 3 can be achieved. However the few results below 1 suggest that it is wise to keep to 1.

Most of the fatigue design rules commonly used across the world were originally based on tests on specimens whose thickness is small compared with many current applications – typically 12 mm thickness – whereas, for example, plate thicknesses of 50–75 mm are not uncommon in offshore platforms and process plant. There is evidence that plate thickness, or its effect on joint configuration and residual stress, influences fatigue crack growth rates and the SN curves in some design guides include means of allowing for this effect. Design data for offshore structures recognise the effect of seawater on fatigue crack growth; for structural areas unprotected from the effects of sea water some require the use of SN curves with reduced life; the threshold or cut-off at low applied stress cycles does not apply, see Fig. 7.17.

The simple SN curve with a cut-off at a certain low stress range derived from constant amplitude data does not reflect the situation with variable amplitude loading when, as was explained earlier, the growing cracking becomes susceptible to smaller and smaller stress ranges. This is allowed for in some guides and standard specifications by altering the horizontal line to

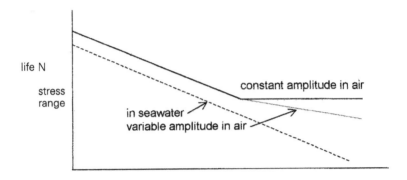

7.17 Constant and variable amplitude design SN curves for various environments.

a sloping line with a slope somewhat less than that of the main part of the line.

Improving on the performance of the as-welded joint

We have seen how the growth of some fatigue cracks in welded joints starts from intrusions at weld toes. It might be reasonable to ask: what would be the situation if these intrusions did not exist? It is easy to find out by removing the intrusions by grinding them out. The SN curve then obtained shows a significant improvement in expected fatigue life for the same applied stress history. What has happened is that the toe of the weld has become plain metal and requires an initiation period of cyclic loading to generate the slip bands and then the vestigial microcracks which grow. There is a further effect in that the rather low cut-off stress for the welded joint is replaced by a fatigue limit as with unwelded material subjected to fluctuating stresses.

The ground welded joint still does not perform as well as plain metal because there are still stress concentrations from the shape of the weld and the joint as well as the effect of residual stresses and the mixed microstructures of the weld. Nonetheless design codes such as are used in the offshore oil and gas industry allow an increase in design life of 120% over life in the as-welded condition. Butt welds can be ground flush but this will not have a significant effect on fatigue life unless it is deep enough to remove the intrusions; there is the possibility that small internal weld flaws near the surface, even porosity, will become surface flaws which could then act as the starting point for fatigue cracks. Grinding can be done with a rotary burr or, more swiftly, with a disc grinder. The latter leaves scratches along the weld and does not give the same level of improvement as the burr which, however, is much slower.

There are other means of improving the fatigue life of welded joints which are perhaps not so readily applied to conventional structures as is grinding. One available for welds in mild and high yield steels is to deform plastically the area around the weld toe with a mechanical peening hammer or needle gun. Shot (not grit) blasting can also be used to achieve a similar effect. These treatments introduce residual compressive stresses in the surface which inhibit the growth of fatigue cracks in addition to which they may modify the toe profile, as with grinding. This type of treatment is not as easily controlled and inspected as is grinding and may introduce damaging effects if the procedure is not correctly designed and proved. Some normal cleaning techniques such as grit blasting or heavy mechanical wire brushing may marginally improve the fatigue life, usually in the sense of improving the worst joints.

Dressing the weld toe with a TIG torch has a strong effect on fatigue life by melting out the intrusions and leaving a smooth toe profile. This is a very slow technique and hardly suited to large constructions but is very cost effective for smaller items. It is of course necessary to design and have qualified a welding procedure for this TIG dressing. Some research work has been done on the effect of paint and other coatings on the fatigue life of welded joints. Such effects as have been shown were attributed to the exclusion of the atmospheric environment from the crack. The tests were usually over a relatively short term and showed some coatings to be more effective than others. However, no paint or non-metallic coating system is totally impervious to the atmosphere and over a long period the benefits do not show up as strongly as in the short term. Of course once larger crack opening does occur, which may be well into the life of the joint, as we saw at the beginning of this chapter, the coating may be unable to sustain the strain and will itself crack, losing any capability to protect the crack tip from the environment.

A traditional technique in some industries for improving fatigue life, the overloading method, has been shown to work on welded joints in steel. Overloading the joint causes yielding in areas which contain residual stresses already at or near the yield. On relaxing the load the residual stress has been reduced and the fatigue performance corresponds more closely to the results expected from a lower stress ratio. The method is relatively powerful because by its nature it attends to the points of highest stress. The method is of course of little or no value where the service stress acts in the opposite sense to the overload, for example, a continuous crane girder, see Fig. 7.18, where the sagging bending moment in the girder in the same bay as the crane may be a hogging moment at the same point when the crane is in an adjacent bay.

Another method which acts on modifying the residual stress pattern is post weld heat treatment or stress relief. This reduces the residual tensile

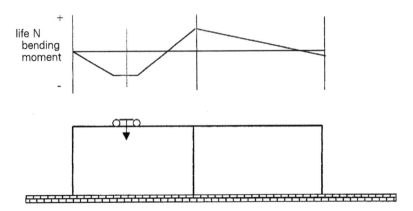

7.18 Crane girder as an example of structure unsuitable for overloading as a fatigue life improvement technique.

stress and again reduces the stress ratio. The cost, time and care required to carry out such a treatment make it hardly worthwhile solely to gain an improvement in fatigue life although the benefits may be taken into account when such treatments are applied for other reasons. The life improvement is small and may just be a manifestation of a reduction in the scatter of the results by improving the worst performing specimens.

When considering these various treatments which can increase the fatigue life it will be seen that most of them involve considerable direct and indirect costs and manufacturing time. Added to this, those relying on treating the local parts of a weld require close quality control; one small missed area or length will render the whole operation ineffective. The increased life derived by surface treatments means that subsurface features then become more significant and so the exploitation of these methods perhaps requires a better class of welding and certainly a more intensive quality control programme. It then has to be said that there is no substitute for designing the as-welded product for the required life in the first place. These treatments are then available for recovering shortcomings or dealing with repairs. Figure 7.19 shows the relative effects of the various life improvement treatments based on test results on one type of detail.

A common feature which can be seen in all of these results is that the life improvement is greater at the longer lives than at the shorter lives. It should be noted that these results were for constant amplitude loading. In variable amplitude loading the higher stresses will tend to modify the residual stress systems set up by peening and overloading which might reduce their benefit. In this type of situation the grinding and TIG dressing are more reliable life improvement techniques. This difference in effect makes it impossible to compare the techniques on a broad basis but Table 7.1 demonstrates the

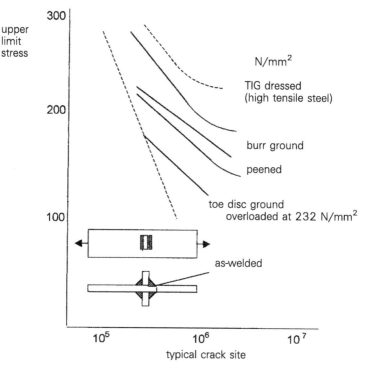

7.19 Comparison of the effect of fatigue life improvement methods for mild steel specimens with transverse non load-carrying fillet welded attachments (after Gurney).

Table 7.1 Effect of improvement techniques

Technique	Life multiplied by maximum of
Painting	3
Toe grinding	8
Shot peening	8
Hammer peening	11
Stress relief	1.5

scale of the effect of the various techniques on a steel plate with a stiffener attached by a butt or fillet weld under a stress range for which the life without treatment would be 2 million cycles.

Using design data

Design data for weld details in terms of fatigue performance referred to earlier in this chapter are given in various publications. For general purpose use one of the most convenient and soundly based sets of fatigue design data for welded joints in steel is BS 7608 *Code of practice for fatigue design and assessment of steel structures*. This categorises various weld details according to their fatigue performance. In essence what the classes do is to allow for the effect of the shape of the joint on the stress in the loaded member so eliminating the need to calculate the stress concentration at the joint. For hollow sections a relevant standard is BS ISO 14347 *Fatigue. Design procedure for welded hollow-section joints*. Analogous fatigue design data for welded aluminium alloys is to be found in publications such as British Standard PD 6702 *Structural use of aluminium. Recommendations for the design of aluminium structures to BS EN 1999*. The design data for each type of detail in BS 7608 is given in the form of SN curves derived from large numbers of laboratory test results which have been resolved into rational data using fracture mechanics parameters and statistical analyses. SN curves representing mean and (mean −2 standard deviations) are given. The mean is the life at which 50% of similar details will fail at a given stress range; this is not really satisfactory for a practical design and so the (mean −2 standard deviations) figure is recommended for most design activities. This gives the life at which there is a 97.5% probability of survival. It can be said that this 'failure', being the complete fracture of a relatively narrow laboratory specimen, is not relevant to a full scale fabrication. However the majority of the fatigue life is spent growing the crack to a size at which it is hardly visible to the naked eye; the difference between the relatively small number of cycles which are then required to fracture the specimen or a full scale item is not significant to the overall fatigue life. It should be borne in mind that such design data are not always suitable for the reverse process of failure analysis except in a relative sense comparing the performance of one type of detail against another. The categories, or class, into which a detail is to be put require the user to identify the type of weld and the joint configuration; what is equally important is that in some classes it is necessary in addition to identify the potential site of the fatigue crack. If this is not done there is the risk of putting the detail into the wrong category. Table 7.2 shows some of the classes against the weld detail and the potential fatigue crack locations. It will be clear that those shown in the table are simple details and that many fabrications are complicated assemblies. It can be quite difficult to attribute a class to weld details in some cases and some application standards contain illustrations to assist in the fatigue classification of welded joints.

There are complicated joints for which this approach is not suitable; in these cases the stress at the toe of the weld has to be calculated. The tubular nodal joint is a widely used example of this. As we saw in Chapter 2, the stress distribution in such joints is the sum of the direct and bending stresses in the individual tubes but with the addition of very high secondary stresses close to the welds due to the local bending of the chord walls. These secondary stresses vary around the line of the joint between brace and chord and

Table 7.2 Classification of weld details by fatigue performance

Description of detail	Class	Explanatory comments	Examples including crack sites
Plain steel			
a) all surface machined and polished	A		
b) in the as-rolled condition, or with cleaned surfaces, but with no flame-cut edges or re-entrant corners	B	Take care not to use Class B for members which may acquire stress concentrations during their life, e.g. rust pitting. In such a case Class C would be more appropriate.	
c) as b) but with any flamecut edges ground or machined to remove visible drag lines	B	Any re-entrant corners in the flamecut edges should have a radius greater than the plate thickness.	
d) as b) but with edges machine flamecut with controlled procedure to ensure that the cut surface is free from cracks	C	Note that the stress used must take into account the effect of any stress concentrations due to the shape of the profile.	
Continuous welds parallel to the applied stress direction			
a) full penetration butt welds fully machined and shown to be free from significant defects by NDT	B	Defect significance should be assessed by fracture mechanics analysis.	
b) butt or fillet welds made by mechanised process with no stop/starts	C	Accidental stop/starts should be repaired to surface and root profile as remainder of weld.	
c) as b) but manual welds or mechanised with stop/start positions.	D		

Table 7.2 (Continued)

Description of detail	Class	Explanatory comments	Examples including crack sites
Transverse butt welds			
a) weld machined flush and proved free from significant defects by NDT	C	Defect significance assessed by fracture mechanics.	
b) as-welded condition with good profile	D	Weld blends smoothly with parent material.	
c) as-welded condition other than b)	E	Applies to welds with 'peaky' profile.	
d) butt weld made on a backing strip without tack welds	F	The crack location is at the root of the weld.	
e) T joint made with butt weld	F	This classification takes account of local bending.	
f) partial penetration butt welds	W on weld throat	Check as full pen butt weld for toe cracking and on weld throat as Class W.	

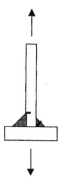

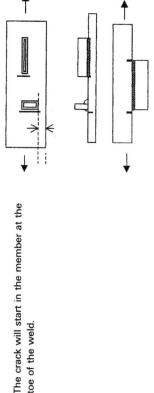

Load carrying fillet welds

a) T or cruciform joints made with fillet welds

F2 with plate stress, W with throat stress

The crack may appear at the weld toe (F2) or in the weld throat (W). The site at which it will appear first will depend on the size of the weld. The joint has to be checked for both sites.

Welded attachments to a stressed member, butt or filet weld

a) attachment within the width of the member, not closer than 10 mm to edge of stressed member

F

The crack will start in the member at the toe of the weld.

b) as a) but on or within 10 mm of edge of stressed member

G

Table 7.2 (Continued)

Description of detail	Class	Explanatory comments	Examples including crack sites
Welds in other locations such as at tubular joints	Basic SN curve T for welds, with hot spot stress	Calculate the highest local stress acting at right angles to the direction of the weld. The diagram is an example only. Depending on the axial/bending ratio the cracking may start at different places around the joint. Use SN curve for application. As an approximation Class D can be used with the calculated local stress.	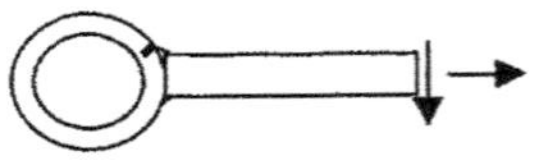
Fillet and partial penetration butt welds in longitudinal shear	W	Use weld throat stress.	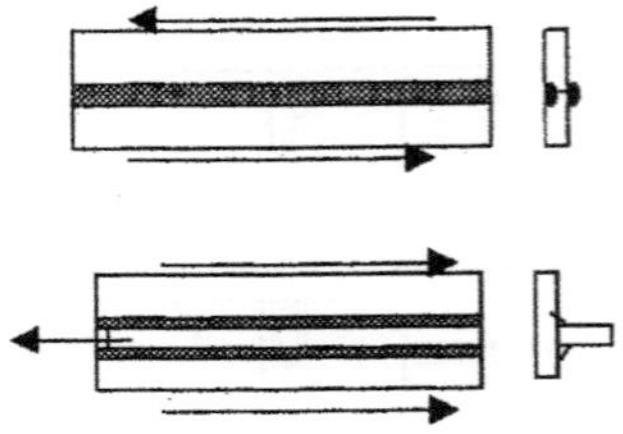

Stress range N/mm²

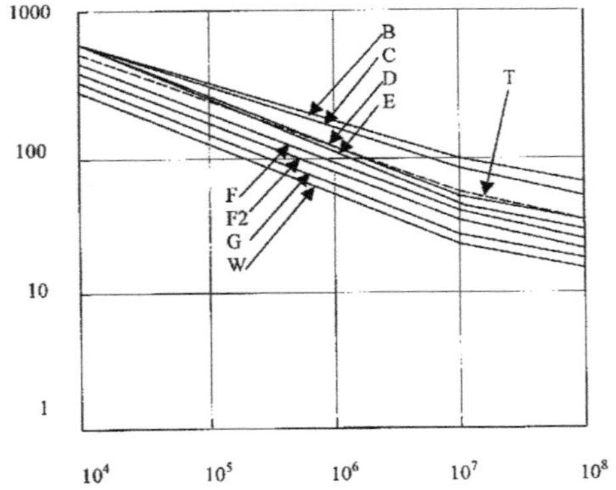

Life in cycles

Typical design SN curves showing welded joint classes.

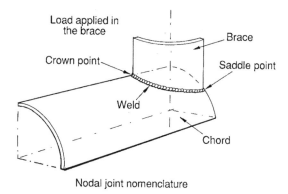

Nodal joint nomenclature

Stress distribution in brace

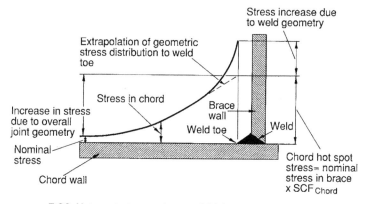

7.20 Hot spot stresses in a nodal joint.

the highest stress is known as the 'hot spot' stress. In a fluctuating load situation this is clearly the stress which will decide the fatigue life of the joint. The design SN curve for this calculation is one derived specially for tubular joints and is in effect the SN curve for the weld on its own; it is therefore no coincidence that it is very close to the SN curve for a transverse butt weld in a flat plate, i.e. Class D. The stress at the weld toe can be calculated either from what are called parametric equations for the particular joint configuration, finite element methods using a computer or by strain measurement on a loaded joint. The latter have to be extrapolated to the weld toe as illustrated in Fig. 7.20.

The motivation for the development of fatigue design data in tubular joints arose mainly from the need for reliability in offshore structures which in the main use circular sections. There is a large scale use of rectangular hollow sections in other industries but these tend not to have such a large element of fatigue loading. As a result there is less design data. Professor Wardenier's book listed in the bibliography discusses this matter within the overall context of hollow section joint design.

Difficulties in deriving the magnitude of the stress range can arise if the principal stress does not act at right angles to the weld toe or if during the load cycle the principal stress changes direction. These circumstances will not be discussed here but are to be found in the work by Maddox in the bibliography.

The following examples of detail classification show how the data can be used.

In the plate girder in Fig. 7.21 the maximum fluctuating tensile stress will be on the extreme fibre on the underside of the lower flange at centre span. There are no weld details on the underside and so that can be classed as B if the edges have been dressed but as C if they are left as flamecut. We then look at the next highest stressed point which is on the top of the lower flange. There are two weld details here, the web to flange weld and the stiffener to flange weld. The web to flange weld is parallel to the direction of stress and so will be a Class D. The stiffener to flange weld is transverse to the direction of stress but does not take any of the flange load and so will be Class F if

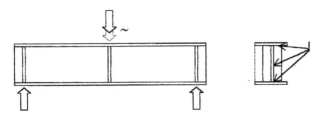

7.21 Plate girder with web stiffeners.

the stiffener is not within 10 mm of the edge of the flange. (If the stiffener is taken to the edge of the flange as shown by the dotted line it will be Class G). In either case the stiffener weld will overrule the web to flange weld because it has a lower class. The stiffener to web weld does not rule because the bending stress dies away from the flange towards the neutral axis. So we are left with Class B or C on the bottom of the lower flange and Class F on top of it. Now the difference in stress between the bottom and top of the flange on this shape of section will probably be very small and the result will be that the Class F will control the design of the flange. So we select the flange size, and so its stress, to give the life we want from a Class F detail. So far we have not considered the size of the web to flange weld; we need to check the shear stress in this weld against Class W. Although the details at the top of the beam are in compression they cannot be ignored, particularly since they are fillet welds. There will be load transfer to the top of the stiffener and web through the web to flange and stiffener to flange welds. We have to make an estimate as to over what length of web to flange weld the load is transferred besides the stiffener to flange welds and then calculate the weld throat stress and ensure that as a Class W detail the life will be adequate.

We can ask if there is any benefit to be had from grinding the toes of the stiffener to lower flange welds. Certainly this will extend the life with respect to toe cracking but we must take account of the fact that the weld is a fillet weld. It can happen that the life is extended by toe grinding but cracking will eventually occur from the root of the weld. This event is not defined by our standard data and so we are unable to say whether it would occur before or after cracks would occur from the web to flange weld. The result is that we could extend the life but we could not say by how much.

We can make a design change to gain a longer life. We do this by eliminating the attachment of the stiffener to the flange, see Fig. 7.22.

The weld around the stiffener then stops in an area of lower bending stress and will give a longer life to the flange. The end of the stiffeners are sniped to give access for welding around the end. It is important that we now check the web stresses, which may be a combination of bending and shear at the end or side of the stiffener to web welds which will be Class F.

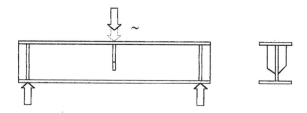

7.22 Plate girder with shortened centre stiffeners.

Summary

- Fatigue cracking is the extension of a crack by each cycle of stress.
- Arc welds have minute intrusions at the toe which can be the starting point for a fatigue crack.
- Fatigue life is proportional to at least the cube of the stress range, half the stress means eight times the life, or more.

Problems

7.1

Suggest four ways of making a butt joint in a tube or pipe and allocate a fatigue class to each.

7.2

This tube is part of a log handling tractor. The lug is the mounting for a hydraulic ram. Fatigue loading is the design case for this item; what type of butt weld should be used at A?

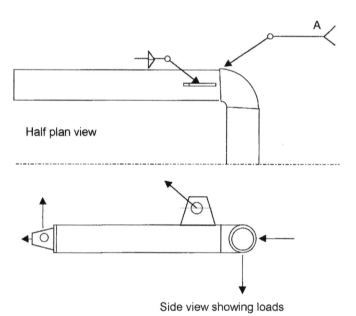

Half plan view

Side view showing loads

8

Brittle fracture

Conventional approaches to design against brittle fracture

A brittle fracture in a metal is a result of crack propagation across crystallographic planes and is frequently associated with little plastic deformation. The propagation of a cleavage crack, as it is known, requires much less energy than does a ductile crack and can occur at an applied stress much lower than that at which failure would normally be expected. In engineering materials such a fracture usually starts from a notch such as a fatigue crack or a welding crack or lack of sidewall fusion; in other words, a high localised stress concentration. A material property which measures the propensity to brittle fracture is the stress intensity at which a brittle fracture occurs. This is known as the critical stress intensity, K_{Ic}. (Stress intensity is described in Chapter 7.) A measure commonly used to verify the resistance of steels to brittle fracture in a production application is the Charpy test (see pages 118–20).

One of the principal reasons why the subject of brittle fracture occupies a key place in the design of steel fabrications is because the ferritic steels, which have a body centred cubic crystal structure, change their fracture behaviour with temperature, from being notch brittle at lower temperatures to being notch ductile at higher temperatures. This is more than an academic distinction because this *transition* from brittle to ductile behaviour takes place close to the ambient temperature of most steel fabrications. The phenomenon is particularly associated with welded fabrications because the energy required to propagate a brittle fracture is low. This means that the stress required to start the crack can be supplied just by the residual stresses from welding without the necessity of an externally applied stress. Furthermore welding can damage the fracture toughness of the steel and in the past some weld metals had very poor fracture toughness. A brittle fracture driven by the strain energy locked up in the metal is a fast moving unstable fracture which has been known to sever complete sections of

8.1 Brittle fracture in ship's hull – M V Kurdistan (photograph by courtesy of TWI).

welded bridges, ships, pressure vessels and pipelines. Figure 8.1 shows a dramatic example. In some cases the crack has been arrested by the exhaustion of the strain energy or by its running into a region of high fracture toughness. The basis of the approach to design and fabrication to resist brittle fracture then lies in appropriate material selection and welding procedure development.

In a limited number of applications steps are taken in design to introduce devices which will arrest a running crack. For example in pipelines the longitudinal welds in adjacent pipe lengths are offset to avoid presenting a continuous path of similar properties along which a fracture could run. As an alternative to this a ring of thicker material or higher toughness material may be inserted at intervals which locally reduces the stress sufficiently to arrest a crack. This is not always effective because in general it requires a material of much higher fracture toughness to stop a crack than would have been necessary to prevent it starting in the first place.

We should also recognise that the consequences of service can also lead to circumstances where a brittle fracture may occur in a fabrication which was initially sound. For example fatigue or corrosion cracks may grow to a critical size during the life of the fabrication and irradiation can reduce the fracture toughness of steels. Materials other than ferritic steels need to have defined fracture toughness but they do not exhibit a significant change of that property with temperature and so the question of material selection has one less dimension. We shall see later in this chapter how the steel can be tested to classify its suitability for use in any particular circumstance but first we need to consider what matters have a bearing on the requirements for fracture toughness.

For any given quality of fabrication these are:

- applied stress
- thickness
- fracture toughness

The criterion of applied stress referred to here is not a question of small differences in calculated stress in a member but whether or not there are large areas of high stress concentration and constraint. Examples of these areas are the nodes in tubular joints where there are large local bending stresses, caused by incompatibility of deformations, and the stress concentrations inherent in openings, nozzles and branches in pipes and pressure vessels. Greater thickness is a feature which engenders tri-axial stress systems which favour plane strain conditions. In addition, thicker material will contain more widely spread residual stress systems than thinner material. For any combination of thickness and stress we can then choose the level of parent metal fracture toughness which research and experience has shown to be appropriate. Perhaps it is not unexpected that the appropriate choice will be set down in a standard specification for the product or application which we have in mind and which itself will refer to a range of steel specifications in another standard. The application will also perhaps introduce as a basis of selection other criteria which have not been mentioned so far such as risk, represented by hazards, their consequences and the likelihood of their occurrence.

Fracture toughness testing and specification

In most steel specifications the measure of fracture toughness is the Charpy test in which a notched bar of the steel is struck by a pendulum. The energy absorbed by the bending and fracturing of the bar is a measure of the fracture toughness of the steel.

Tests are done on a number of samples at different temperatures and the energy absorbed is found to vary with temperature. The change of energy occurs over a range of temperature called the *transition temperature range*, see Fig. 8.2. The energy measured is not a fundamental measurement which can be mathematically related to quantities such as stress intensity although certain empirical relationships have been derived. However as a result of experience certain minimum values of Charpy test energy have been found

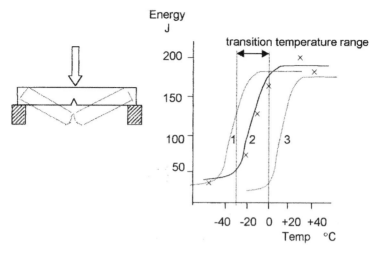

8.2 Charpy test specimen and typical results.

which give freedom from brittle fracture in conventionally fabricated constructions.

In Fig. 8.2 the full line, 2, is the curve given by the set of results marked X. The higher results to the right are on what is called the *upper shelf* although the minimum values required by many specifications will often be found in the transition range but above the *lower shelf* figures. The steelmaker can produce steels with different fracture toughnesses and different transition ranges as in lines 1 and 3. Within the carbon manganese steels this is done by a combination of metallurgy, mechanical working and heat treatments. Generally the finer the grain size of the steel and the fewer the non-metallic inclusions the higher will be the fracture toughness. This property in the parent material determines the lowest temperature at which a fabrication can be used, provided that it is not overridden by the weld and heat affected zone properties. The minimum temperature at which it is practical to use carbon manganese steel fabrications is around –40°C. The alloy steels containing around 9% nickel are suitable down to around –190°C. Below that temperature austenitic steels or aluminium alloys can be used but their fracture toughness still has to be controlled.

The temperature at which the minimum Charpy energy is specified is not necessarily the minimum temperature at which the fabrication can be used. The Charpy test specimen is of a standard size, 55 × 10 × 10 mm, regardless of the thickness of the steel from which it is taken. The effects of thickness which we have considered mean that as the thickness goes up we have to use a steel with the minimum required Charpy energy at lower test temperatures. For example a typical offshore platform specification requires a certain minimum energy level at the Charpy test temperatures shown below for as-welded, i.e. not post weld heat treated, fabrications in regions of high stress. These regions would normally be the nodal joints in tubular structures.

Thickness (mm)	Charpy test temperature (°C)
≤20	−20
>20 and ≤100	−40

For regions which are not highly stressed the Charpy test temperature for the 20–100 mm thickness is set at −30°C. If the fabrication is heat treated after welding, i.e. thermally stress relieved, the Charpy test temperatures for this thickness range can be increased to −30°C for high stress regions and −20°C for other parts, and for non-high stress regions less than 20 mm in thickness, −10°C. These requirements are based on a minimum design temperature of −10°C. If a different design temperature is used then the Charpy test temperature should be changed by 0.7°C for every 1°C that the design temperature differs from −10°C. This is a somewhat arbitrary method of fixing a test temperature and is based on an assumed shape of the transition curve. These temperatures apply to carbon-manganese steels of all strengths and a different Charpy energy is required of each grade of steel, typically the minimum energy required is equivalent numerically in Joules to one tenth of the highest minimum yield strength of that grade of steel in N/mm^2. This is necessary because the energy required to bend a Charpy test specimen prior to fracture in a higher yield strength steel will be greater than that required for a lower strength steel.

Other products such as buildings and bridges have their own require-ments which are usually less demanding than those for offshore construc-tion; they recognise the service conditions, the consequences of failure and the customary levels of control in the respective industries. These requirements are expressed in various ways. In some products the required Charpy test temperature for the steel is related to a range of thicknesses. In others the thickness of the steel requires a certain grade of steel without direct reference to a Charpy value or some other measure of notch toughness.

Fracture mechanics and other tests

Where a more discriminating test, or one giving results which can be applied to the assessment of defects, is required a fracture mechanics test can be used. Such tests can use a specimen from the full thickness of the material under study and with a crack starting notch which is more representative of actual weld defects than the rather blunt notch of the Charpy specimen.

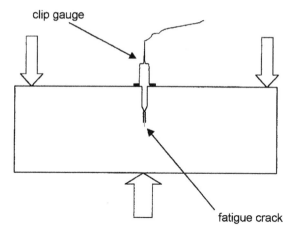

clip gauge

fatigue crack

8.3 Specimen set up for crack tip opening displacement test.

In Chapter 7 we saw that the state of stress around the tip of a sharp crack can be described by a quantity known as the stress intensity. In a fully elastic material this quantity may reach a critical value at which fracture occurs, K_{Ic}. We can measure this by carrying out a fracture mechanics test which entails bending a cracked specimen and measuring the load at which fracture occurs. By calculating the stress at the crack tip at fracture the value of K_{Ic} can be calculated. This can be used to make an assessment of the significance and acceptability of weld defects or fatigue cracks.

However, we have seen that structural steels are far from being elastic when they reach yield point. A result of this is that even at low applied stresses the crack tip actually stretches plastically and this can be measured as the crack tip opening displacement (CTOD), δ. The value of this as measured at fracture is used in assessment of the significance of cracks or other features, particularly in welded joints. The measurement of CTOD is made on a notched bar loaded in bending. In preparing the specimen, see Fig. 8.3, the notch is first sawn and then grown by fatigue cracking to produce the finest possible crack tip. The opening of the notch is measured by an electrical displacement gauge and the actual tip opening is calculated on the basis of the crack and bar geometry.

The use of a fatigue cracked notch not only ensures that the finest crack is produced but it can be placed within a welded joint so as to sample quite narrow regions of a particular microstructure in the weld metal or the heat affected zone.

In Fig. 8.4, a 'K' preparation has been used to give a heat affected zone straight across the section so that the fracture will always be within the same microstructure as it moves into the specimen.

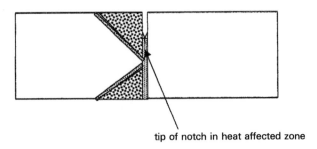

tip of notch in heat affected zone

8.4 CTOD test specimen for a butt weld showing the tip of the notch in a selected region of microstructure.

Summary

In order to minimise the risk of brittle fracture in service, the following steps should be taken:

- Select steels which have a defined weldability and fracture toughness properties suited to the service requirements.
- Employ fabrication and welding procedures and welding consumables which have the proven capability to achieve the required properties.
- Define weld flaw acceptance levels which the steel and weld metal in the thickness to be used will tolerate under the design stress and temperature.
- Operate a management system which will ensure that the procedures are followed and will achieve the required quality of fabrication.

Assessment of structural integrity

Defect assessment

There are a number of sophisticated analytical and measurement techniques which enable engineers with the relevant detailed knowledge and experience to predict whether the presence of some feature in a structure represents a hazard to the integrity or the life of the structure of which it is a part. The results of these techniques can be used to set general product *acceptance standards* or to assess the effect of some particular feature of a product – a defect or flaw – which does not meet the general product standard. There can be no other branch of engineering in which so much time and money has been spent and so much written on trying to demonstrate that defective manufacturing is acceptable as in the welded fabrication industry. In the early days of welding when the composition of materials, the knowledge of their behaviour and the performance of personnel and equipment was somewhat erratic there might have been some justification. However, enormous strides have been made in recent years in all these areas and there is no longer any excuse or reason for not producing arc welded joints whose homogeneity and properties match those of the parent metal.

However, in reality we appear not to have reached that position or at least industry has no confidence that it has. There are of course more constructive uses of the techniques such as in assessing the integrity or residual life of items which have been damaged by wear and tear or accident. The procedures employed can be extremely complicated and their full exploitation requires a deep knowledge of the inherent assumptions and the corrections which have to be made to allow for specific circumstances.

Various fractures in structures as varied as bridges, ships, power generation and chemical plant and pipelines have been studied over the years and have been the subject of assessments of the significance of defects which have enabled the methods to be developed, verified and improved.

The systematic evaluation of the residual life of fatigue cracked structures has been practised in the aircraft industry for many years. Until the late

1950s the *safe life* approach was used whereby a structural component was allowed a certain number of flying hours based on tests or calculation with a suitable factor of safety built in. When these hours were about to be exceeded the component would be replaced. This was unsatisfactory from two points of view: firstly, some components failed before their allowed life and secondly, some, if left in place, could have safely exceeded their allowed life by a large margin. In the first case human tragedy would strike as well as financial loss; in the second case a costly component would have been unnecessarily scrapped. In the 1950s steps were taken to avoid these extremes by recording the load histories of individual aircraft so that replacement of components could be timed to correspond to the actual fatigue damage and not just flying hours. The recording of load histories also gave benefits in that operators could monitor the fatigue life and take steps to devise operating procedures and flight plans which minimised fatigue damage thereby extending the life of their aircraft and reducing the cost of maintenance.

This safe life approach was eventually replaced by what was known as a *fail safe* approach. This was made feasible by the introduction of multiple load paths (structural redundancy) and crack tolerant materials with crack stopping features where redundancy was not achievable. As with many terms, 'fail safe' does not really fit the concept as it works. It is really a 'partial-failure-safe' or a 'crack-safe' approach in which one load path or part of one can be lost by cracking without reducing the short term integrity of the structure. The introduction of this approach into wing design accompanied the replacement of wing designs in which the bending load was taken by a single spar, as an I beam, by a design in which the bending load was taken by the wing skin, as a box girder, in which the skin had to be thicker than previously and could only be made by machining from solid slabs, see, Fig. 9.1.

The skin was made up of a number of 'planks' which accounted for the structural redundancy and the number of shear webs was increased to match these planks which in essence gave a multi-box structure. The essence of this change of design philosophy was that it required, or promoted, whichever way you look at it, a completely new manufacturing method, one was

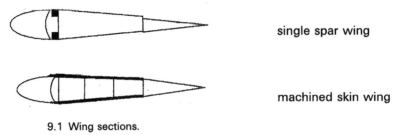

single spar wing

machined skin wing

9.1 Wing sections.

impossible without the other. In assessment terms the concept was relatively simple to apply since the wing skin was thin compared with the bulky spar sections and only two-dimensional crack growth prediction was required in most places. The technique was also viable because by then load histories of civil aircraft were reasonably well recorded. Once a crack was detected the remaining life of the cracked component until repair or replacement became necessary could be predicted in terms of operating hours and plans made for repairs to be done at the next scheduled overhaul. Other aspects of aircraft residual life were not so readily addressed, in particular corrosion, which is less amenable to prediction and relies more on inspection – possibly requiring extensive stripping and dismantling.

Analogous techniques are applied to many types of industrial plant to examine the possibility of extending the working life of the plant as an alternative to heavy capital investment in new plant. An overall review is made of the condition of the plant and its historic, current and future operating conditions. New knowledge and improved methods of predicting such mechanisms as crack growth, creep, wear and corrosion are employed to arrive at an estimate of the future life of the plant which may be shown to be longer than was originally projected, sometimes without any modification at all.

British Standard BS 7910 'Guide to methods for assessing the acceptability of flaws in fusion welded structures' comprehensively describes some of these techniques. These can be very complicated and require substantial knowledge and experience of the techniques to be able to select the appropriate level and judge what aspects of the structure, material and its behaviour must be allowed for. Two simple applications of the techniques will be outlined here; this will show what type of information is required to be able to make an assessment. It should not give every welding engineer the impression that he can become a confident practitioner in the techniques. Unless the welding engineer has experience of such assessments it is best to seek the advice of a specialist engineer experienced in the derivation and practice of the methods. The Guide addresses fracture and plastic collapse, fatigue, creep and creep fatigue, leakage, corrosion, erosion, corrosion fatigue, stress corrosion and instability as they may arise from the presence of flaws as represented by cracks, lack of fusion or penetration, cavities, inclusions and shape imperfections. For assessment with respect to fracture the publication advises that the application of the procedures should be performed by 'competent personnel with practical and theoretical experience of the type of structure to be assessed and the methods described'. It offers a choice of three levels of assessment depending on the input data available, the level of conservatism and the degree of complexity required. The Level 1 procedure has been designed 'for practising engineers with a peripheral knowledge of fracture toughness concepts and stress analysis', which does not sound as demanding as the earlier 'competent personnel . . .' statement. More extensive knowledge is required for Level 2, whereas Level 3 applications will generally require a deep knowledge of fracture mechanics concepts. The reader will have to decide into which of these categories he falls before deciding which way to move if faced with a situation which appears to deserve attention. Assessment with respect to fatigue does not have a set of levels but offers a choice between a general and a simplified procedure. Assessment in respect of creep is offered at several levels of complexity depending on the criticality of the problem and the material property data available. Other modes of failure are addressed in briefer and more general terms.

Fracture assessment

We will take as an example a planar flaw such as a crack appearing at the surface of a structural member, see Fig. 9.2, and whose plane is at right angles to the surface. It will probably have an irregular shape when viewed on its face.

The mathematics cannot deal with this irregular shape and so the crack is idealised as a rectangle of the size containing the crack shape as shown in Fig. 9.2. In this example the crack is intended to be of a size which does not significantly reduce the cross sectional area. The fracture toughness used for the material in which the crack lies is derived either from tests giving K or δ. The values of these quantities should be those relevant to the type of fracture behaviour being considered. For this example a straightforward brittle fracture in a structural steel is to be used and the values will be K_{Ic} or δ_c and the Level 1 procedure is chosen.

As we saw in Chapter 7 the applied stress intensity factor has the general form

$$K_I = (Y\sigma) \sqrt{\pi a} \qquad [9.1]$$

where $(Y\sigma)$ is the stress which would be in the material in the absence of the crack allowing for stress concentration effects which may include weld shape, overall joint configuration, misalignment, etc. together with corrections for finite width and bulging where these are relevant. For

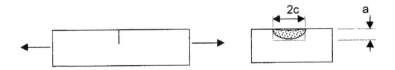

9.2 Example of crack in loaded member.

as-welded structures, allowance for residual stress can be made where appropriate to the location of the crack with respect to the weld.

Solutions for a wide variety of geometries are given in the Guide.

The applied CTOD, δ_I, is found from the following equation using K_I as defined above when, for steels the total stress, σ_{max}, i.e. the sum of all the stresses acting where the crack lies, is less than 50% of the yield stress

$$\delta_I = \frac{K_I^2}{\sigma_y E} \qquad [9.2]$$

Where the total stress is greater or equal to 50% of the yield stress

$$\delta_I = \frac{K_I^2}{\sigma_y E} \left[\frac{\sigma_y}{\sigma_{max}} \right]^2 \left[\frac{\sigma_{max}}{\sigma_y} - 0.25 \right] \qquad [9.3]$$

By this stage the welding engineer will perhaps begin to realise why this type of calculation is best left to the specialist but nonetheless we will continue if only to show what information is required to be able to undertake an assessment.

Having acquired the K and δ we then have to calculate the *fracture ratio* which is the ratio of the applied stress intensity to the critical stress intensity or in terms of CTOD, the ratio of the applied CTOD to the critical CTOD. These ratios are given by

$$K_r = \frac{K_I}{K_{Ic}} \qquad [9.4]$$

and

$$\sqrt{\delta_r} = \sqrt{\frac{\delta_I}{\delta_c}} \qquad [9.5]$$

The Level 1 assessment is a simplified method of checking if a certain flaw gives rise to a risk of fracture. The flaw sizes derived have an inherent safety factor on size of around 2. A flaw is deemed acceptable if K_r is less than $1/\sqrt{2}$, i.e. less than 0.707. The opposite procedure of finding the flaw size which is acceptable is pursued by defining a flaw size parameter, $\bar{a}$, as the half length of a through thickness crack, see Fig. 9.3.

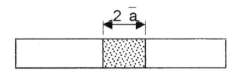

9.3 Through thickness crack.

The maximum tolerable value of $\bar{a}$ is $\bar{a}_m$ which is defined by the following expressions as appropriate

$$\bar{a}_m = \frac{1}{2\pi} \left[\frac{K_{Ic}}{\sigma_{max}} \right]^2 \tag{9.6}$$

or

$$\bar{a}_m = \frac{\delta_c E}{2\pi \left[\dfrac{\sigma_{max}}{\sigma_y} \right]^2 \sigma_y} \tag{9.7}$$

Or where the total stress is greater than or equal to 50% of yield stress

$$\bar{a}_m = \frac{\delta_c E}{2\pi \left[\dfrac{\sigma_{max}}{\sigma_y} - 0.25 \right]^2 \sigma_y} \tag{9.8}$$

The Guide includes charts which show how $\bar{a}$ can be derived for various shapes of crack.

Fatigue assessment

In welded steel structures the assumption remains, as in Chapter 7, that there are yield point stresses close to the weld so that if we are assessing a crack in a weld the superimposed applied stress effectively pulsates downwards from yield strength irrespective of applied stress ratio. No allowance is made for stress relief treatments because they are considered to leave a substantial level of residual stress in place.

Two methods of assessing weld flaws with respect to fatigue are offered. The fracture mechanics method asks the user to integrate a chosen crack growth rate expression up to the maximum allowable crack size which may be a limit set by such occurrences as exceedance of yield or tensile strength, brittle fracture, leakage, instability or resonance.

For all values of ΔK above a threshold value, ΔK_0, the crack growth law is

$$\frac{da}{dN} = A(\Delta K)^m \tag{9.9}$$

For ΔK less than ΔK_0, da/dN is assumed to be zero. Values of A and m depend on material and applied conditions and can be taken as constant over a limited range of ΔK. Values of A and m for steels in ambient air and other environments are published and may be found in the Guide. For steels, including austenitic steels, the recommended values of the constants for steel in air up to 100°C are:

$$m = 3$$
$$A = 5.21 \times 10^{-13}$$
$$\Delta K_0 = 63$$

For da/dN in mm/cycle and ΔK in N/mm$^{3/2}$.

The value of ΔK is taken from one of the published solutions in the Guide or elsewhere. The crack is then 'grown' incrementally until it reaches a predetermined shape or size. If in growing, the crack changes its form, for example, from being a buried crack to a surface breaking crack, it will have to be treated as the new form from there on. If a variable stress history has to be used then consideration must be given to the order in which the various stress ranges are introduced. If the small ones are introduced at the beginning the threshold value will eliminate their damage and if the large ones are introduced too early on in the history their potential effects on the larger crack sizes will be underestimated if that is where they occur in practice. A cycle by cycle procedure is clearly rather tedious even by computer and it may be more convenient to divide the stresses into blocks which are large enough to reduce the work but small enough to avoid the problem of sequencing referred to above. Another suggested method is to postulate a certain crack growth increment and work 'backwards' to calculate the number of cycles required to reach that point.

It will be appreciated that this integration method is really rather cumbersome. The simpler quality category method therefore has great appeal for more routine use. Like many such methods its description is more complicated than its execution. What is given in the Guide is a set of SN curves each of which is allocated to a certain arbitrary weld quality level in steels. Some of these SN curves are the same as those allocated to standard weld details in documents such as BS 7608. The flaw dimensions are used to define the initial parameter $\bar{a}_i$ and the parameter corresponding to the maximum size to which the flaw will be allowed to grow, $\bar{a}_m$. The values of $\bar{a}$ are then to be entered on a chart which gives a notional stress range figure which is then set against a quality level. If this is higher than that equivalent to the welded joint in question then the flaw is acceptable.

Summary

- Most structures are designed to standard requirements of strength and quality.
- It is possible to go beyond the bounds of the standard requirement by undertaking an engineering critical assessment of the structure or one joint.
- In dealing with fracture and cracking a sound knowledge of the basis of fracture mechanics is required.

1.1

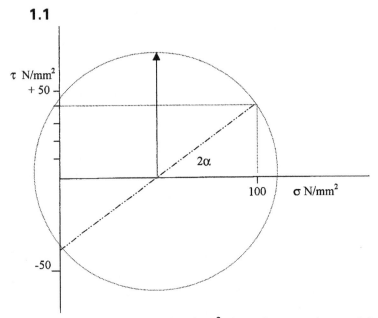

Plot the stresses that we know, 100 N/mm² along the normal stress (σ) axis and 40 N/mm² up the shear stress (τ) axis. Then 0 N/mm² on the normal stress axis (there is no normal stress in the vertical direction) and –40 N/mm² down the shear stress axis, the complementary shear stress acting at right angles to the applied shear stress. We then draw the circle through the two points which we have plotted of which the radius is then by simple geometry:

$$\sqrt{40^2 + 50^2} = 64$$

The angle $2\alpha = \tan^{-1} 40/50 = 38.7°$

And so reading off the diagram the maximum principal stresses are +114 N/mm² at an angle, α, of 19.3° to the longitudinal axis of the beam and –14 N/mm² at right angles to that. The maximum shear stress is 64 N/mm².

1.2

1 bar = 0.1 N/mm^2

The hoop stress is $pd/2t$ = 1 × 600/2 × 3 = 100 N/mm^2
and the longitudinal stress is $pd/4t$ = 50 N/mm^2
There are no applied shear stresses in the hoop or longitudinal directions
and so we can plot the stresses 100 and 50 N/mm^2 on the normal stress axis.

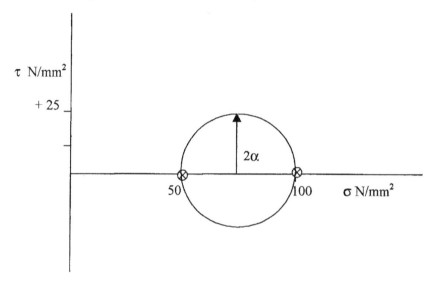

The maximum shear stress is then seen to be (100 −50)/2 = 25 N/mm^2 at an
angle of 90/2 = 45° to the longitudinal axis, i.e. along the weld.

2.1

At the front end there is not much bending moment or shear force in most conventional aircraft configurations, unlike the centre section just at the back of the wing which has the maximum bending moment and shear as well as torque from the fin and rudder. Take a look at this back end of the fuselage on the ground and you will see that many of the skin panels buckle under the gravity (1g) load.

So right at the front with 1 bar = $0.1 \ \text{N/mm}^2$
the hoop stress is $pd/2t = 0.6 \times 0.1 \times 6000/2t < 150$

So the skin thickness, t, might be 1.2 mm

2.2

The cyclist's weight is entirely on the saddle, one hand is just touching the handlebars and his feet are not touching the pedals. Ignore the slight offset of the cyclist's weight from the apex of the frame at the saddle pillar.

$$P_R + P_F = 75$$

Taking moments for the whole frame about the rear wheel hub:

$$75 \times 400 = P_F \times 1000 \quad P_F = 30 \ \text{kg} \quad P_R = 75 - 30 = 45 \ \text{kg}$$

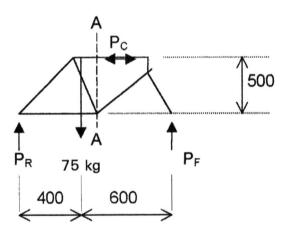

Take moments about the bottom bracket, to the right of section AA:
$30 \times 600 = 500 \ P_c \quad P_c = 36 \ \text{kg}$ which is approximately 360 N.

6.1

The torque, T, on the tube is 6 kNm. The arm of the weld is the tube radius $d/2$, 37.5 mm, and the weld length is πd, $\pi 75$ mm, on each side of the plate. We could be more fussy and say that the centroid of the fillet weld is at a greater radius than $d/2$ but the difference in the final answer will be small.

Weld torque $= 2\pi d \times d/2 \times t\tau$ where t is the weld throat and τ is the allowable weld throat stress

So $6 \times 1000 \times 1000 = 2\pi 75 \times 37.5 \times 120 \times t$

From which $t = 2.8$ mm corresponding to a leg length of 4 mm.

7.1

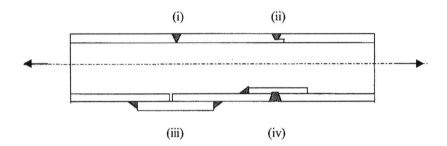

(i)   is a full penetration butt weld with full root penetration and a smooth
      underbead and cap. This will be a Class D detail.
(ii) is a spigot joint, in fatigue terms this appears to be the same as a butt
      weld on a backing, Class F, but note that the weld throat does not equal
      the wall thickness and so the stress should be based on the actual weld
      throat.
(iii) are load carrying fillet welds and so the weld throat stress should be
      used with a Class W SN curve.
(iv) is a butt weld on a permanent backing. With the tack weld as shown
      (which will be feasible only on large diameter tubes) the detail is Class
      G but if the backing is not tacked inside the tube then it will be Class F.

7.2

We might begin by saying that we need the best fatigue performance and so
a full penetration butt weld is necessary, perhaps with a TIG root run and a
consumable insert to get the best root profile.

However, the weld of the lug to the tube is a fillet weld which is at one
end close to the joint 'A'. This means that at this point the tube stress will
have to be kept down to Class F at best so there is no point in going for a
high performance type of butt weld at virtually the same point. So what butt
weld is equivalent to a Class F SN curve? The answer to that is a butt weld
on a permanent backing strip. The strip must be a good fit inside the pipe
bore, perhaps a sprung one is a good idea here. If we wanted a more
cautious approach we might use a ceramic backing strip which, if it gave a
smooth root profile, could give a Class D or E performance.

The fillet weld is a strange choice for the lug attachment when for not
much extra cost a full penetration T-butt weld could be made. The fillet weld
would be Class W on the weld throat stress and F on the lug plate itself. A
full penetration weld would remove the risk of throat failure still leaving the
plate as Class F, as will be the tube.

Chapter 1　Fundamentals of the strength of materials

Case J, Chilver Lord and Ross C T F, *Strength of materials and structures*, 3rd edition, London, Edward Arnold, 1993.

Roark R J and Young W C, *Roark's Formulas for stress and strain*, New York, McGraw-Hill, 1989.

Timoshenko S, *Strength of materials, Part I*, New York, D Van Nostrand, 1955.

Chapter 2　Stresses in some common types of structures

Gill S S (ed), *The stress analysis of pressure vessels and pressure vessel components*, Oxford, Pergamon Press, 1970.

Marshall P W, *Design of welded tubular connections*, Amsterdam, Elsevier, 1992.

Wardenier J, *Hollow section joints*, Delft, Delft University Press, 1982.

Significance of deviations from design shape, London, Institution of Mechanical Engineers Pressure Vessels Section and Applied Mechanics Group, 1979.

Steel designers' manual, 5th edition, Oxford, Blackwell Scientific Publications, 1992.

National structural steelwork specification for building construction, London, British Constructional Steelwork Association, 1994.

Chapter 3　Elementary theories of bending and torsion

Case J, Chilver Lord and Ross C T F, *Strength of materials and structures*, 3rd edition, London, Edward Arnold, 1993.

Roark R J and Young W C, *Roark's Formulas for stress and strain*, New York, McGraw-Hill, 1989.

Timoshenko S, *Strength of materials, Part I*, New York, D Van Nostrand, 1955.

Chapter 4   Basis of design of welded structures

Davies J M & Brown B A, *Plastic design to BS 5950*, Oxford, Blackwell Science, 1996.

Specification for unfired fusion welded pressure vessels. BS 5500, London, British Standards Institution.

Eurocode 3. *Design of steel structures. BS EN 1993*. (Various Parts)

Chapter 5   Weld design

Weld symbols on drawings. Abington, Abington Publishing, 1982.

Welding and allied processes. Recommendations for joint preparations. Manual metal arc welding, gas shielded metal arc welding, gas welding, TIG welding and beam welding of steels. ISO 9692, Geneva, International Organisation for Standardization.

Welded, brazed and soldered joints. Symbolic representation on drawings. BS EN 22553: 1995. London, British Standards Institution, ISO 2553 'Welded, brazed and soldered joints – Symbolic representation on drawings'. Geneva, International Organisation for Standardisation.

Welding terms and symbols. BS 499, London, British Standards Institution.

Welding, brazing and soldering processes – Vocabulary. ISO 857, Geneva, International Organisation for Standardization, 1990.

Chapter 6   Calculating weld size

Design rules for arc welded connections in steel. Doc. XV-358-74, International Institute of Welding, 1974 (unpublished).

International test series. Doc. XV-242-68, International Institute of Welding, 1968 (unpublished).

Chapter 7   Fatigue cracking

Gurney T R, *Fatigue of welded structures*, Cambridge, Cambridge University Press, 2nd edition, 1979.

Hobbacher A (ed), *Fatigue design of welded joints and components*, Cambridge, Abington Publishing, 1996.

Maddox S J, *Fatigue strength of welded structures*, Cambridge, Abington Publishing, 2nd edition, 1991.

Niemi E (ed), *Stress determination for fatigue analysis of welded components*, Cambridge, Abington Publishing, 1995.

Richards K G, *Fatigue strength of welded structures*, Cambridge, Abington Publishing, 1969.

Wardenier J, *Hollow section joints*, Delft, Delft University Press, 1982.

Chapter 8   Brittle fracture

Boyd G M (ed), *Brittle fracture in steel structures*, London, Butterworths, 1970.

Honeycombe R W K, *Steels – microstructure and properties*, London, Edward Arnold, 1981.

Chapter 9   Assessment of structural integrity

Guide to methods for assessing the acceptability of flaws in metallic structures. BS 7910: 1998, London, British Standards Institution, 1998.

The following are works on subjects mainly outside the immediate scope of this book but providing valuable welding and materials background.

Bailey N, *Weldability of ferritic steels*, Cambridge, Abington Publishing, 1994.

Burgess N T (ed), *Quality assurance of welded construction*, London, Elsevier Applied Science, 2nd edition, 1989.

Croft D, *Heat treatment of welded steel structures*, Cambridge, Abington Publishing, 1996.

Davies A C, *The science and practice of welding*, vols 1 & 2, Cambridge, Cambridge University Press, 10th edition, 1992.

Gibson S and Smith A, *Basic welding*, Basingstoke, MacMillan, 1993.

Gourd L M, *Principles of welding technology*, London, Edward Arnold, 1980.

Higgins R A, *Materials for the engineering technician*, 3rd edition, London, Arnold, 1998.

Lancaster J F, *Metallurgy of welding*, Cambridge, Abington Publishing, 6th edition, 1999.

Lancaster J F, *Handbook of structural welding*, Abington, Abington Publishing, 1997.

Welding and allied processes – Nomenclature of processes and reference numbers. ISO 4063, Geneva, International Organisation for Standardisation.

Index

Printed and bound by CPI Group (UK) Ltd, Croydon, CR0 4YY

03/10/2024

01040848-0019